Medical Intelligence Unit

Protein-Based Inheritance

Yury O. Chernoff, Ph.D.
Georgia Institute of Technology
Atlanta, Georgia, U.S.A.

CRC Press
Taylor & Francis Group
Boca Raton London New York

CRC Press is an imprint of the
Taylor & Francis Group, an **informa** business

Protein-Based Inheritance

Medical Intelligence Unit

CRC Press
Taylor & Francis Group
6000 Broken Sound Parkway NW, Suite 300
Boca Raton, FL 33487-2742

First issued in hardback 2019

ISBN 13: 978-1-58706-138-7 (hbk)

Visit the Taylor & Francis Web site at
http://www.taylorandfrancis.com

and the CRC Press Web site at
http://www.crcpress.com

Library of Congress Cataloging-in-Publication Data

A C.I.P. Catalog record for this book is available from the Library of Congress.

About the Editor...

YURY O. CHERNOFF is a Professor at the School of Biology, Georgia Institute of Technology, Atlanta, Georgia (U.S.A.). He received his Ph.D. in Biology from St. Petersburg State University (Russia), and performed postdoctoral research at Okayama University (Japan) and University of Illinois at Chicago (U.S.A.). His main research interests include yeast prions and other yeast-based models for protein aggregation disorders, with specific emphasis on the cellular control of protein aggregates and on the evolution of prion properties. His work provided the first experimental evidence for the chaperone role in prion phenomena. Dr. Chernoff is Editor-in-Chief of the journal *Prion* and serves on the Editorial Board of *Gene Expression*.

Dedication

Dedicated to people who have enough strength and bravery to stand their ground even against a majority of others in the endless pursuit of truth.

CONTENTS

EDITOR

Yury O. Chernoff
Georgia Institute of Technology
Atlanta, Georgia, U.S.A.
Email: yury.chernoff@biology.gatech.edu
Chapters 7,8

CONTRIBUTORS

Ilya M. Alexandrov
Cardiology Research Center
Moscow, Russia
Chapter 6

Janine Beisson
Centre de Génétique Moléculaire
CNRS
Gif-sur-Yvette, France
Email: janine.beisson@cgm.cnrs-gif.fr
Chapter 9

Lee Byrne
Research School of Biosciences
Univeristy of Kent
Canterbury, Kent, U.K.
Chapter 5

Brian S. Cox
Protein Science Group
Department of Biosciences
University of Kent
Canterbury, Kent, U.K.
Chapters 2,5

Irina L. Derkatch
Department of Microbiology
New York University School of Medicine
New York, New York, U.S.A.
Email: irina.derkatch@med.nyu.edu
Chapter 4

Herman K. Edskes
Laboratory of Biochemistry and Genetics
NIDDK, NIH
Bethesda, Maryland, U.S.A.
Chapter 1

Abbi Engel
Laboratory of Biochemistry and Genetics
NIDDK, NIH
Bethesda, Maryland, U.S.A.
Chapter 1

Sergey G. Inge-Vechtomov
Department of Genetics
St. Petersburg State University
and
St. Petersburg Branch of Vavilov Institute of General Genetics
Russian Academy of Sciences
St. Petersburg, Russia
Chapter 8

Dmitry Kryndushkin
Laboratory of Biochemistry and Genetics
NIDDK, NIH
Bethesda, Maryland, U.S.A.
Chapter 1

Vitaly V. Kushnirov
Cardiology Research Center
Moscow, Russia
Chapter 6

Susan W. Liebman
Department of Biological Sciences
University of Illinois at Chicago
Chicago, Illinois, U.S.A.
Chapter 4

Linsay McCann
Laboratory of Biochemistry and Genetics
NIDDK, NIH
Bethesda, Maryland, U.S.A.
Chapter 1

Toru Nakayashiki
Laboratory of Biochemistry and Genetics
NIDDK, NIH
Bethesda, Maryland, U.S.A.
Chapter 1

Eugene G. Rikhvanov
Siberian Institute of Plant Physiology
and Biochemistry
Siberian Division
Russian Academy of Sciences
Irkutsk, Russia
Chapter 7

Nina V. Romanova
School of Biology and Institute
for Bioengineering and Bioscience
Georgia Institute of Technology
Atlanta, Georgia, U.S.A.
Chapter 7

Sven J. Saupe
Laboratoire de Génétique Moléculaire
des Champignons
UMR 5095 CNRS
Université de Bordeaux 2
Bordeaux, France
Email: sven.saupe@ibgc.u-bordeaux2.fr
Chapter 3

Frank Shewmaker
Laboratory of Biochemistry and Genetics
NIDDK, NIH
Bethesda, Maryland, U.S.A.
Chapter 1

Michael D. Ter-Avanesyan
Russian Cardiology Center
Moscow, Russia
Email: mdter@cardio.ru
Chapter 6

Mick F. Tuite
Department of Biosciences
University of Kent
Canterbury, Kent, U.K.
Email: m.f.tuite@ukc.ac.uk
Chapters 2,5

Aleksandra B. Vishnevskaya
Cardiology Research Center
Moscow, Russia
Chapter 6

Reed B. Wickner
Laboratory of Biochemistry and Genetics
NIDDK, NIH
Bethesda, Maryland, U.S.A.
Email: wickner@helix.nih.gov
Chapter 1

Patricia G. Wilson
Regenerative Bioscience Center
Department of Animal
and Dairy Science
University of Georgia
Athens, Georgia, U.S.A.
Email: pgwilson2@earthlink.net
Chapter 10

Galina A. Zhouravleva
Department of Genetics
St. Petersburg State University
Petersburg, Russia
Chapter 8

PREFACE

This book covers a topic which has been neglected for years and has come back into the spotlight only recently. Until the genetic role of DNA was firmly established, many researchers suspected that proteins rather than nucleic acids could be carriers of heritable information. However, these models were completely forgotten with the triumphal march of the double helix and development of a central dogma, postulating that information flow occurs strictly from DNA through RNA to protein, making it seemingly impossible for the proteins to possess a coding potential. Proteins were downgraded to the role of simple perpetuators and executors of DNA orders. While it was certainly recognized that protein "serfs" are indispensable for the well-being of their powerful nucleic acid "lords", the thought of a protein occupying a key position in the hereditary hierarchy was as unthinkable in modern molecular genetics as was the peasant's to the king's throne in medieval Europe.

As aspiration frequently occurs in science, data that could not be explained within the framework of a "nucleic acid only" model of heredity existed for years and were just waiting for the proper moment to resurface. Attention to these non-conventional phenomena focused on the transmissible spongiform encephalopathies (TSEs), later termed "prion diseases". Accumulated results led to the model proposing that a TSE infectious agent (prion) is composed of the wrongly shaped protein, capable of converting the normal protein of the same amino acid sequence into a prion shape. As usual, this revolutionary model was not immediately accepted by the scientific community. Yet it has eventually gained recognition, highlighted by a Nobel Prize awarded to S. Prusiner for studying prion diseases in 1997. Despite this, some researchers remain skeptical in regard to the "protein only" nature of the TSE agent even today. Mammalian prion diseases were covered in great detail in some recent books (for example, *Prion Biology and Diseases* Cold Spring Harbor Laboratory Press, Cold Spring Harbor, NY, 2004, edited by S. Prusiner, and *Prions and Prion Diseases: Current Perspectives*, Horizon Scientific Press, 2004, edited by G. Telling). Although transmission of mammalian prions from one organism to another represents infection rather than inheritance, the ability of a protein to act as an information carrier, postulated in the prion model of TSEs, certainly paved the way for acceptance of such a role in heredity for at least some proteins.

Application of the prion model to some yeast non-Mendelian elements was introduced by R. Wickner in 1994 and confirmed by further research in various labs. Prions of yeast and other fungi manifest themselves as protein-based heritable elements, thus demonstrating that proteins may serve as carriers of hereditary information. Most fungal prions are heritable amyloids (that is, ordered fibrous aggregates) propagated by nucleated polymerization, that is similar to a mechanism proposed for mammalian prion diseases. Moreover, recent advances leave no doubts that patterns of the fungal prions are controlled and reproduced exclusively at the protein level. Prion (or prion-like) phenomena based on self-activating protease and possible self-activating kinase were also described. In principle, prion inheritance does not contradict the central dogma as DNA still remains needed for initial protein production. However, prion transmission in generations clearly shows that changes occurring at the protein level can become reproducible and heritable without any corresponding change occurring in DNA sequence. Heritable prions essentially meet all the criteria of non-Mendelian genes that were in use during the classic genetics era. Connection between mammalian and fungal prions, as well as genesis of the yeast prion concept and some structural and biological implications drawn from recent studies of fungal prions are described by R. Wickner and coauthors in Chapter 1.

The prion model provides a molecular basis for some fungal non-Mendelian phenomena that were studied for years but were not be understood within the framework of conventional concepts. Chapters 2 (by M. Tuite and B. Cox) and 3 (by S. Saupe) provide a comprehensive description of two such phenomena, respectively [*PSI*$^+$] of *Saccharomyces cerevisiae* and [Het-s] of *Podospora anserina*, with inclusion of both historical recollections and current data explaining the mechanisms of previously observed patterns. Studying various combinations of the amyloid-based prions in yeast uncovered complex interactions between them, in some cases leading to generation of prion "cascades", where pre-existing prion facilitates de novo formation of the prion isoform of an unrelated protein. These and other types of prion-prion interactions are described in Chapter 4 by S. Liebman and I. Derkatch.

Although most of the known prions (that is, infectious or heritable proteins) are amyloids, there are other amyloids (including those associated with important human diseases) which do not possess prion properties. Specific features of the yeast prion proteins, that make these proteins capable of propagating the prion state and distinguish them from non-heritable amyloids, are considered in Chapters 5 (by B. Cox and coauthors) and 6 (by M. Ter-Avanesyan and coauthors). A crucial distinction is a pattern of interactions between the amyloid aggregates and certain chaperone proteins that play a major role in propagation of yeast prions in vivo. Chapter 7 (by E. Rikhvanov and coauthors) compares the effects of chaperones on prion and non-prion aggregates, concluding that these effects are based on the same molecular mechanisms, but lead to different consequences depending on the nature of the protein substrates. As a result, the chaperone machinery designed to protect cells from protein aggregates is turned by a prion into a tool for propagation of prion aggregates.

Despite a significant amount of information accumulated during recent years, the biological role of prion phenomena remains a matter of extensive discussions. Chapter 8 (by S. Inge-Vechtomov and coauthors) considers potential biological roles of prion domains in detail, describing both the pathological and potentially adaptive effects of amyloid formation and pointing to mechanisms that may govern the evolution of prion-forming sequences.

While fungal prions surely represent the most extensively studied example of protein-based inheritance, it is not the only example known to date. Ironically, the oldest proven case of the hereditary change which does not involve DNA is almost as old as the DNA-centered concept of inheritance. This first evidence for protein-based inheritance has come from research in Protozoa, specifically in ciliates, where it has been shown that a surgically produced alteration of the complex multiprotein structure becomes heritable in a template-like fashion. This eventually led to development of the "structural templating" concept, also applicable to prion phenomena. These and other data on the inheritance of preformed structures in Protozoa are considered in Chapter 9, written by one of the pioneers of this research, J. Beisson.

Finally, Chapter 10 by P. Wilson deals with the long debated phenomenon of centrosome inheritance which was considered as a potential example of non-chromosomal templating. Even though recent data do not support protein-based templating being involved in the centrosome reproduction, they certainly reveal certain effects of pre-existing structures on the formation of new centrosomes during cell division. Examples described in Chapters 9 and 10 touch on "the chicken and the egg" issue...the origin and reproduction of the eukaryotic cytoskeletal structures in general, a matter that is far from being solved.

Taken together, data included in this book prove beyond a reasonable doubt that proteins and multiprotein complexes are able to control heritable traits, and that at least in some examples, this control occurs in a template-like fashion, so that new structures strictly reproduce patterns of pre-existing structures that were not specifically coded in DNA. Thus, protein-based inheritance has left the area of speculation and has emerged as a new topic amenable to high-quality experimental analysis. Nucleic acid lords will no longer be capable of disregarding the contributions of their protein serfs to the overall heritable composition of the cell and organism. Moreover, connections between the mechanisms of protein-based heritable phenomena and some important diseases (such as Alzheimer's disease and other disorders related to amyloid formation) make it probable that protein-based inheritance will attract even more attention in the near future.

Certainly this book represents only the tip of the iceberg, and its composition is biased due to both the state of the field at the moment and the preferences of the editor, that can never be put aside completely, no matter how hard one tries to do so. If presentation of the proven cases of protein-based inheritance weakens the natural "DNA chauvinism" of the readers, while discussion of emerging speculations leads some researchers to explore alternative explanations of the other mysterious phenomena, then I could conclude that the book has played its role as intended. This is not to say that knowledge of DNA sequences of everyone and everything is unimportant. This is just to note that it may not be enough. And while exact knowledge of the firmly established facts is certainly rewarding, this is a call of the unknown that always drives science forward.

Remaining fascinated and obsessed with this call, and being grateful beyond words to the great team of authors who actually wrote this book, I offer the product of our work to the readers.

Yury O. Chernoff, Ph.D.

Acknowledgements

I would like to thank all the people who made this book possible, beginning with our Publisher, Ron Landes, who approached me with the initial idea of producing such a book. I am proud of our great team of authors, who were willing to contribute their time and effort to writing this book. I am also grateful to our great team of reviewers, whose names I promised to keep hidden, but whose input was essential nevertheless. I also thank the Landes Bioscience staff, including Cynthia Conomos, Celeste Carlton and Bonnelle Martin, who directed and guided the process of book preparation from beginning to end. This book would never be completed without the dedicated members of my lab who helped me to keep the research process going even when I had to concentrate on writing and editing. Finally, I would like to thank my family for endless patience and support.

CHAPTER 1

Yeast Prions:

Evolution of the Prion Concept

Reed B. Wickner,* Herman K. Edskes, Frank Shewmaker, Toru Nakayashiki, Abbi Engel, Linsay McCann and Dmitry Kryndushkin

Abstract

Prions (infectious proteins) analogous to the scrapie agent have been identified in *Saccharomyces cerevisiae* and *Podospora anserina* based on their special genetic character istics. Each is a protein acting as a gene, much like nucleic acids have been shown to act as enzymes. The [URE3], [*PSI*+], [*PIN*+] and [Het-s] prions are self-propagating amyloids of Ure2p, Sup35p, Rnq1p and the HET-s protein, respectively. The [β] and [C] prions are enzymes whose precursor activation requires their own active form. [URE3] and [*PSI*+] are clearly diseases, while [Het-s] and [β] carry out normal cell functions. Surprisingly, the prion domains of Ure2p and Sup35p can be randomized without loss of ability to become a prion. Thus amino acid content and not sequence determine these prions. Shuffleability also suggests amyloids with a parallel in-register β-sheet structure.

The Genesis of the Prion Concept from Studies in Mammals

The transmissible spongiform encephalopathies (TSEs) of mammals are inexorably fatal degenerative brain diseases whose etiology has long been debated,[1,2] but are widely believed to be caused by an infectious protein. The unusual radiation-resistance of the scrapie agent[3] generated a flurry of speculation on its nature, including a surprisingly accurate early version of the protein-only hypothesis.[4] It was proposed that an altered form of a cellular protein binds a monomer of the normal form, and in this complex, changes the normal to the abnormal form. This is, in essence, the modern view. The key protein was identified genetically as the *Sinc* gene of mice controlling scrapie incubation period.[5] However, it was only 18 years later that *Sinc* was shown to be the gene encoding PrP,[6] the major component of the scrapie agent.[7]

PrP is a nonessential protein[8] located on the cell surface where it is bound by a GPI anchor.[9] Animals lacking the *Prnp* gene encoding PrP are immune to infection by the TSE agent,[10] showing neither pathology, nor substantial replication of infectivity. PrP from brains of TSE-infected animals is quite protease resistant, compared to the protease-sensistive normal protein. It accumulates significantly in diseased tissues because of reduced turnover. The precise structure of the TSE-form of PrP (called PrP-res or PrP^{Sc}) is not known, but it is clearly higher in β-sheet content than the normal protein. Amyloid deposits composed largely of PrP-res are observed in many but not all TSEs. The smallest infectious material is estimated to be a 14 to 28-mer, but most of the infectivity is much larger.[11] The protease-resistance of infectious material also suggests that it is amyloid in form, even if frank plaques are not always seen.

*Corresponding Author: Reed B. Wickner—8, Room 225, NIH, 8 Center Drive MSC 0830, Bethesda, Maryland 20892-0830, U.S.A. Email: wickner@helix.nih.gov

Protein-Based Inheritance, edited by Yury O. Chernoff. ©2007 Landes Bioscience.

While extensive circumstantial evidence points to the TSEs being prion diseases, with the infectious agent nothing more than an altered PrP, definitive experiments are still not available, and there continues to be some debate on this point. The best evidence to date comes from studies in which amyloid formed in vitro from recombinant mouse PrP89-230 was injected into mice transgenic for PrP89-231. The mice developed a scrapie-like disease, albeit after an inordinately long incubation period, and their brains were infectious for normal mice.[12] The fact that this oft-attempted experiment has so far only worked with amyloid of PrP again indicates that amyloid is indeed the infectious material.

The TSEs are infectious, hereditary and spontaneous. Brain extracts of one animal will readily infect another animal on injection or ingestion. Human hereditary Creutzfeldt-Jakob disease is caused by mutations in the gene for PrP, presumably making it more likely to spontaneously assume the prion form. Spontaneous cases are presumed due to spontaneous formation of infectious amyloid by the normal PrP protein.

Injection of infected brain extract into brains of uninfected animals produces disease with a long, but very characteristic incubation period. The incubation period is much longer for infections across species lines (the 'species barrier'). Distinct TSE strains (or variants) have been defined, with different incubation periods, distinguishable signs and symptoms and biochemical characteristics of the altered PrP. These strain (or variant) characteristics are not due to different PrP sequences, but are thought to reflect different amyloid structures. The TSE strain (variant) also affects the species barrier: while one TSE strain may be unable to cross between a particular pair of species, another may readily do so (reviewed in ref. 13).

Discovery of Infectious Proteins (Prions) in *S. cerevisiae*

When yeast is supplied with a good nitrogen source, such as ammonia, it turns off transcription of the genes encoding the enzymes and transporters (e.g., *DAL5*, encoding the allantoate transporter) needed to use poor nitrogen sources, like proline or allantoate (reviewed in refs. 14,15). This control mechanism is called nitrogen catabolite repression or nitrogen control and is mediated by Ure2p. [URE3] is a nonchromosomal gene whose dominant effect is to derepress these enzymes and transporters.[16] [PSI^+] is a nonchromosomal gene discovered as a translational suppressor of nonsense mutations,[17] and Sup35p is a subunit of the translational termination factor.[18,19] The molecular basis of [URE3] and [PSI^+] was long a puzzle.

We proposed three genetic criteria to distinguish nucleic acid replicons such as viruses and plasmids from prions[20] (Fig. 1).

Reversible Curability

While a virus, plasmid or prion may be curable (efficiently eliminated) by some treatment, a virus or plasmid is not likely to be regenerated de novo in less than geologic time. However, the protein capable of becoming a prion is still present in the cured strain and could spontaneously convert to the self-propagating altered prion form.

Overexpression of the Protein Increases Frequency of Prion Generation

Overproducing a chromosomally encoded protein will not increase the frequency with which a plasmid or virus arises de novo, but increasing the cellular content of a protein able to become a prion should increase the frequency of prion generation. The change must be self-propagating, and so should take over most of the population of molecules of that protein, converting them to the prion form.

Phenotype Relation and Gene Dependence

For viruses, plasmids and prions, the propagation of the nonchromosomal element always requires the activity of some chromosomal proteins. Prion propagation requires at least the gene encoding the protein. If the prion form of a protein were simply an inactive form of the normal protein, then the phenotype of the prion-carrying strains should resemble that of

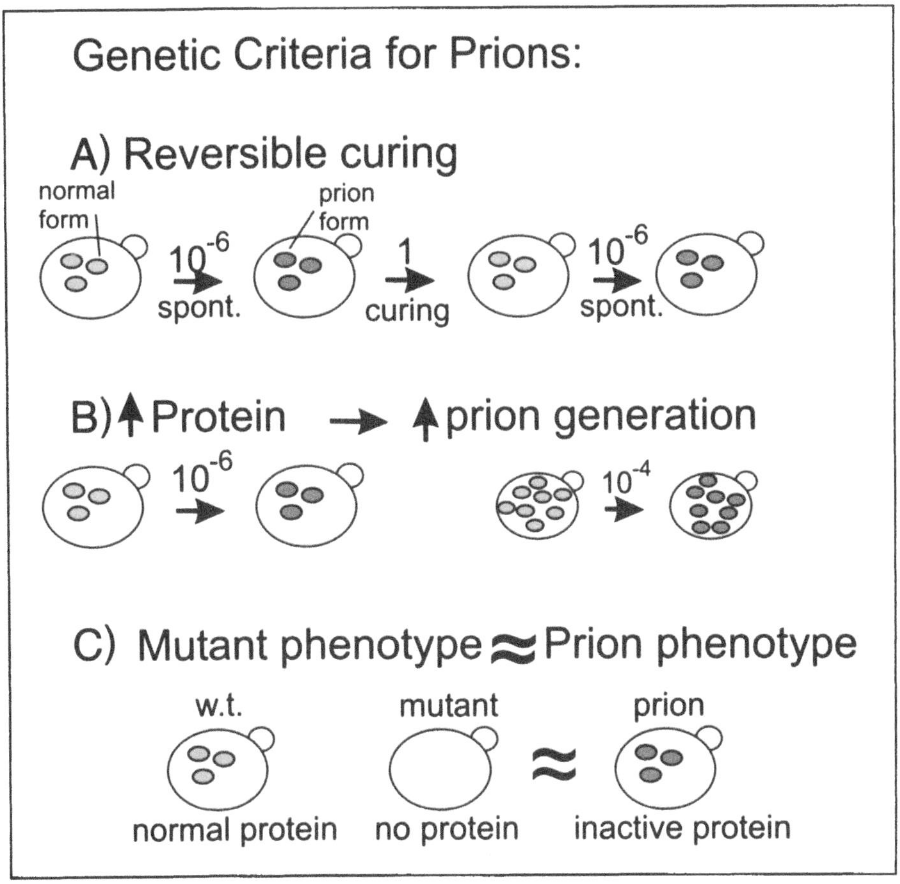

Figure 1. Genetic criteria for prions. Reversible curing means that in a strain cured of a nonchromosomal genetic element, the same element can arise again. Overproducing a protein with the potential to become a prion increases the frequency with which the prion arises. If the prion form of the protein is an inactive form of the protein, then the phenotype of the presence of the prion is the same or similar to that of a mutant in the gene for the protein. Each of these three properties should be characteristic of prions but none of them are known (or expected) for nucleic acid replicons such as plasmids or viruses.

mutants in the gene encoding the protein. This contrasts with viruses or plasmids conferring a cellular phenotype (such as the mitochondrial DNA or killer virus). In these cases, the phenotype of mutation of a chromosomal gene needed for propagation of the nucleic acid replicon is that of absence of the replicon (e.g., killer-negative, glycerol minus).

[URE3] and [PSI^+] Are Prions

[URE3] has all of these properties if viewed as a prion of Ure2p, and [PSI^+] qualifies as a prion of Sup35p.[20] [URE3] can be cured by guanidine but arises again at a low frequency.[20] The overproduction of Ure2p elevates the frequency of [URE3] by 20 to 200 fold.[20] Finally, the phenotype of [URE3] strains is very similar to that of *ure2* mutants.[21] [PSI^+] may be cured by high osmotic strength,[22] but [PSI^+] derivatives of the cured strains are easily isolated.[23] Overproduction of Sup35p elevates the frequency of [PSI^+] arising de novo,[24] and the [PSI^+] phenotype resembles that of *sup35* mutants, namely, nonsense suppression.[17]

What Does It Take to Be a Prion?

We found that the N-terminal asparagine-rich part of Ure2p was necessary and sufficient to propagate and induce the [URE3] prion,[25,26] and at the same time we reinterpreted the similar results of TerAvanesyan et al on the Q/N rich N-terminal domain of Sup35p.[27] We call these the prion domains of the respective proteins. It is clear that mutations within the prion domain can affect prion propagation.[28-31] Some of these changes do not prevent the protein from being a prion, but rather introduce a 'species barrier' between molecules.[32,33]

To examine whether there are sequence determinants of prion-formation ability, the prion domains of Ure2p and Sup35p were each shuffled, leaving amino acid content and codon usage unchanged.[34,35] Five shuffled variants of each prion domain were generated and reintroduced into the chromosome in place of the normal prion domain (Fig. 2). It was found that each of the shuffled prion domains of Ure2p and of Sup35p were capable of being prions. The frequency of prion formation varied somewhat and in each case, one in five of the shuffled sequences produced only unstable prion variants. But all could be prions.[34,35] In addition, each of the shuffled Ure2p species readily formed amyloid in vitro. In support of this picture, a detailed deletion analysis of the Ure2p prion domain showed that no single region of the prion domain is essential for prion-forming ability.[35]

These results imply that the composition of the prion domain is the critical determinant of prion formation. It is very likely that the high Q/N content of the Ure2p and Sup35p prion domains is important. However, few of the many proteins with such Q/N-rich domains have been found capable of making prions. There are doubtless other compositional features of the prion domains that are important. Their relatively low content of charged residues and hydrophobic amino acids are probably important, but further work will be needed to define the critical features.

Because small deletions in the C-terminal domains of Ure2p[25] and larger deletions of Sup35p[36] C-terminus dramatically increase the frequency of prion formation, it was suggested that the prion domain and C-terminal domains interact, preventing the prion domains from interacting with eachother to form amyloid. However, no evidence for such an interaction could be detected,[37] and the Ure2p prion domain appears to be unstructured in its native (soluble) form. The fact that the prion domain can be shuffled and still support prion formation and propagation argues that if there is such an interaction, it is not important for this process.

Shuffleable Prion Domains Suggest Parallel In-Register Structure

Although amyloids have long been known to be rich in β-sheet, their more detailed architecture has been unclear. There are at least three mutually exclusive possibilities for the β-sheet architecture of amyloid. An antiparallel β-sheet has adjacent strands bonded to each other running in opposite orientations: N->C next to C->N, for example the amyloid of the Aβ(34-42) fragment.[38] In a parallel in-register β-sheet structure, adjacent bonded strands are in the same orientation: N->C next to N->C, and **identical** residues are bonded to each other, for example the amyloid of Aβ(1-40).[39-42] Electron spin resonance indicates that amyloids of amylin and of α-synuclein also have parallel in-register β-sheet structure.[43,44] A third possibility is some form of parallel out-of-register β-sheet, for example the β-helix structure of pectate lyases.[45] Here, like the antiparallel structure, **nonidentical** residues are paired.

Amyloid formation is much like a linear crystal, in that essentially a single species of protein is singled out to join the growing filaments. This specificity demands that there be some specificity in the bonding between chains. For anti-parallel β-sheets or β-helices, this could be large with small, positive with negative, hydrophobic with hydrophobic, hydrogen bonding (donor) with hydrogen bonding (recipient). In these cases, shuffling the sequence would disturb the alignment of complementary residues, and presumably prevent prion formation (Fig. 3).

For parallel in-register β-sheets, hydrogen bonding between Q/N residues[46] or S/T residues, or hydrophobic with hydrophobic residues could provide specificity. However, charged

Shuffled prion domains can still become prions

Ure2p

Sequence	Prion	Amyloid
Wild type: MMNNNGNQVSNLSNALRQVNIGNRNSNTTTDQSNINFEFSTGVNNNNNNN SSSNNNNVQNNNSGRNGSQNNDNENNIKNTLEQHRQQQQ	+	+
Scrambled sequence #1: MVDGNQMNNNKSRRNSSQRGNSNQRVNNQNENNFNGLAQSSNNNNSITTT FTNNNQINSQLNGINNNVNQTDQNVQNHGNSNENNSENL	+	+
Scrambled sequence #2: MQSHQAESNSSQNGDQNGTNNLQNNRSNGINNFGNNNRNQNNLESQRVNN TINNNKLNQFNGNNEVNNVQNQSSDNTNNNMSIVTTRNS	+	+
Scrambled sequence #3: MNIRNQNQSTAVLNVNQQSNNGTSNSVNNLNFNNSGMQNHGRNFNQSTRN NNTNEKGGNNILNSNDERINNQQNQENNNTVDNSQNNSS	+	+
Scrambled sequence #4: MMQRNGQQEGTNNNHSNINTQRNVFNNSANNNRNNNEGLNNNNSNFNNLV SNNQQVNVSSNSNINNQDNNKSILSGTSNDTTENRGQQQ	+/-	+
Scrambled sequence #5: MNTNNSQGSFVDENQNRSIVKSRTVNMSQNNNTGNNNNAQLNNILNNTDS GHVSNNENRLGRQNNDFNQNSSQTNNGNNQQQSNNNNNI	+	+

Sup35p

	Sequence	Prion formation
wild type	MSDSNQGNNQQNYQQYSQNGNQQQGNNRYQGYQAYNAQAQ PAGGYYQNYQ GYSGYQQGGYQQYN PDAGYQQQYN PQGGYQQYN PQGGYQQQFN PQGG RGNYKNFNYNNNLQGYQ .	+
Scrambled		
seq. 1	MSQYNSYNGQQQQFQNNQYQGNRDQQGYSGYQGDAYNQQ GQYYPQGNGPNYYNYGNQGYKNQQAGQQGGYANNGFNGL NYRYGPQYQGYYNYNQGQPPQNAYQAQQQGQQNQSN...	+
seq. 2	MSNNQNGQPNNPYRQNGQQYYYNPQNSYGANGQQNRYGY YQNQYYQQAGGSNQFGNQGYDKNGYQYQFQQAQYQNQLY QGYANNGNSGQQNAGNPYGGGGYQDQQQGQQYYQQP...	+
seq. 3	MSSAQQKNNQAGQQQNQPQQGYGSNRQQYQNSQQPYYGG NGYYQGQYPGGGYFYQGNNNQNQPFQQNPDLYGQGQYGY AGYNYGNNANANGQQNQDGGYYQYNNYQQYNQQQRY...	+
seq. 4	MSYQYQNNYGNQGNQGQQYNNYSQGGAQPQLQRYQYQQP YPQYYYGNQNQQYYQQYQPQQYGGANYRGQNQNFYGNDQ YAGQYGGGNNNNGSGDQGNSNAKPQGQQNGQFNQYA...	+
seq.5	MSYQGYQNGQYYQNKFYDQNGSYQGSGYNNNQYGQQQQY NYQQAGGYNYANQYGNPNQRPGQGNGQQSQQQQYQRPQY NQYYQAQNNQPQGFQNDGYNQGGGNANGAPNYGLYQ...	+

Figure 2. Scrambled prion domains can still be prions.[34,35] In place of the normal Ure2 or Sup35 prion domains, shuffled prion domains (five of each) with the same amino acid content were constructed and integrated. Each of the shuffled prion domains could be a prion, although one of each was unstable.

residues (which are few in the prion domains of Ure2p and Sup35p) should tend to interfere with formation of this structure. Shuffling the residues of a parallel in-register β-sheet does not change the pairing, since identical residues are always paired. Thus, we argue that if a prion domain can be shuffled and still be a prion, it should have a parallel in-register β-sheet structure.[47]

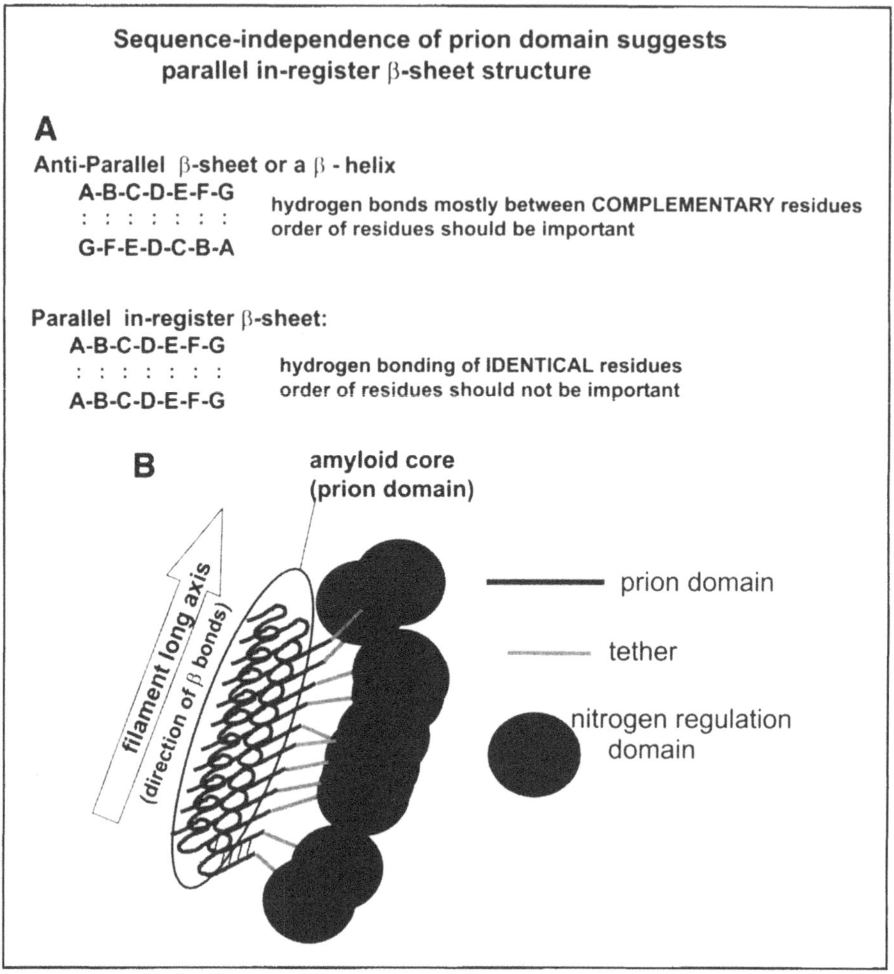

Figure 3. A prion domain insensitive to scrambling should be a parallel in-register amyloid.[47] A) Nonidentical residues are bonded in an anti-parallel β-sheet or β-helix. The specificity of amyloid propagation (similar to crystal growth) implies that there must be some complementarity of residues. Shuffling such a sequence would destroy any such complementarity and thus prion formation. Shuffling a parallel in-register β-sheet leaves identical residues paired with eachother. If a prion domain can be shuffled and not lose prion-forming ability, it suggests a parallel in-register β-sheet structure. B) Model of Ure2p amyloid structure (see text). A color version of this figure is available online at www.eurekah.com.

This suggests the sort of model shown in Figure 3B. The core of the amyloid is made up of Ure2p1-65,[48,49] which should be a parallel in-register β-sheet.[47] Indeed Ure2p10-39, a fragment of the prion domain, has been shown to have such an architecture.[50] The folding of the β-sheet is demanded by the diameter of the amyloid filaments of the prion domain.[48] Ure2p66-95 is unstructured in both native and amyloid forms of Ure2p, and we call this the 'tether' (green in Fig. 3B). The C-terminal part of Ure2p apparently does not change its conformation on formation of amyloid.[51,52] A similar parallel in-register β-sheet model can be proposed for Sup35p, since its prion domain is shuffleable and the charged M domain is likely to serve as a tether.

Amyloid Is the Prion Infectious Material, Not a Dead End (Side-) Product

In a ground-breaking study, infection of *Podospora anserina* with the [Het-s] prion by amyloid of recombinant HET-s protein was acheived.[53] Soluble protein was not infectious nor was heat- or acid- denatured aggregated protein. The transmissibility of [*PSI*⁺] by amyloid of recombinant Sup35p has also been demonstrated, and evidence was also obtained that the amyloid structure determines the prion variant.[54,55] As mentioned above, amyloid of recombinant PrP has also shown some infectivity for mice.[12]

We have now demonstrated the ability of amyloid formed in vitro from recombinant Ure2p to infect cells with the [URE3] prion[56] (Fig. 4). The low level infectivity of soluble Ure2p (Fig. 4B) is apparently due to filament formation while the experiment is in progress. Cells infected with amyloid of recombinant Ure2p show at least three prion variants, distinguishable by their mitotic stability and by the intensity of their phenotype (degree of *DAL5* derepression). Extracts of [URE3] strains are also infectious, and transmit the [URE3] variant that was present in the strain from which the extract was prepared (Fig. 4C). Remarkably, the amyloid made in vitro from recombinant Ure2p is as much as 1/3 as infectious as is an extract (on a per Ure2p molecule basis).[56] The extracts can be used to seed amyloid formation by soluble recombinant Ure2p, but the extent to which this amplification is variant-faithful is limited by the tendency of the 'soluble' Ure2p to spontaneously form amyloid filaments, the latter having a mixture of variant structures.[56]

The Ure2p prion domain by itself, or fused to various other proteins can also form amyloid which is infectious.[56] Cells infected with these fusion proteins (or prion domain alone) show the same spectrum of prion variants as those infected with amyloid formed from the full length protein.

Preliminary size fractionation experiments indicate that infectious material is greater than 20 nm in diameter, indicating a filament length of >40 mer. Amyloid filaments must be sonicated to be infectious, apparently in order to get into yeast. However, while the largest size fraction of filaments has only low infectivity, resonication increases its infectivity many fold.[56] We suggest that this increase in infectivity is a combination of generation of new filament ends (which must be the growing point) and of allowing more facile entry into the cells.

The infectivity of amyloid (and not soluble or other aggregated forms) in all of the prion systems indicates that amyloid is not a dead-end or side product of the prion process. The structure of amyloid formed in vitro has long been recognized to be morphologically heterogeneous. Recently evidence for structural heterogeneity of Aβ amyloid has been obtained.[57] It is clear from the prion studies that prion variants are encoded by differences in amyloid structure. It will be particularly interesting to know what are these structural differences and how they propagate.

[*PSI*⁺] and [URE3] Are Diseases of Yeast

It has been proposed, based on plate tests, that [*PSI*⁺] is an advantage to cells carrying it in surviving stress[58] and for evolvability.[59,60] Some strains grow better under certain conditions if they are [*PSI*⁺] than if they are [*psi*⁻], although there are no conditions that uniformly favor [*PSI*⁺], and most conditions favor [*psi*⁻].[58-60] The genetic basis for these phenotypes remains to be determined.

All of the conditions were measuring growth, but yeast may be spending most of its time in stationary phase. To what extent are the few conditions favoring [*PSI*⁺] represented in the wild? This question is almost impossible to answer directly, and it is further complicated by the fact that whether [*PSI*⁺] is favored or unfavored is very strain-dependent.

We examined the distribution of [URE3], [*PSI*⁺] and [*PIN*⁺] in 70 wild strains.[61] Prions arise de novo and spread by infection, so that even if they are a mild disadvantage to their host, they should be frequently found in the wild. As controls, we examined the distribution of parasitic DNA and RNA replicons of yeast: the 2 micron DNA plasmid, 20S RNA, 23S RNA

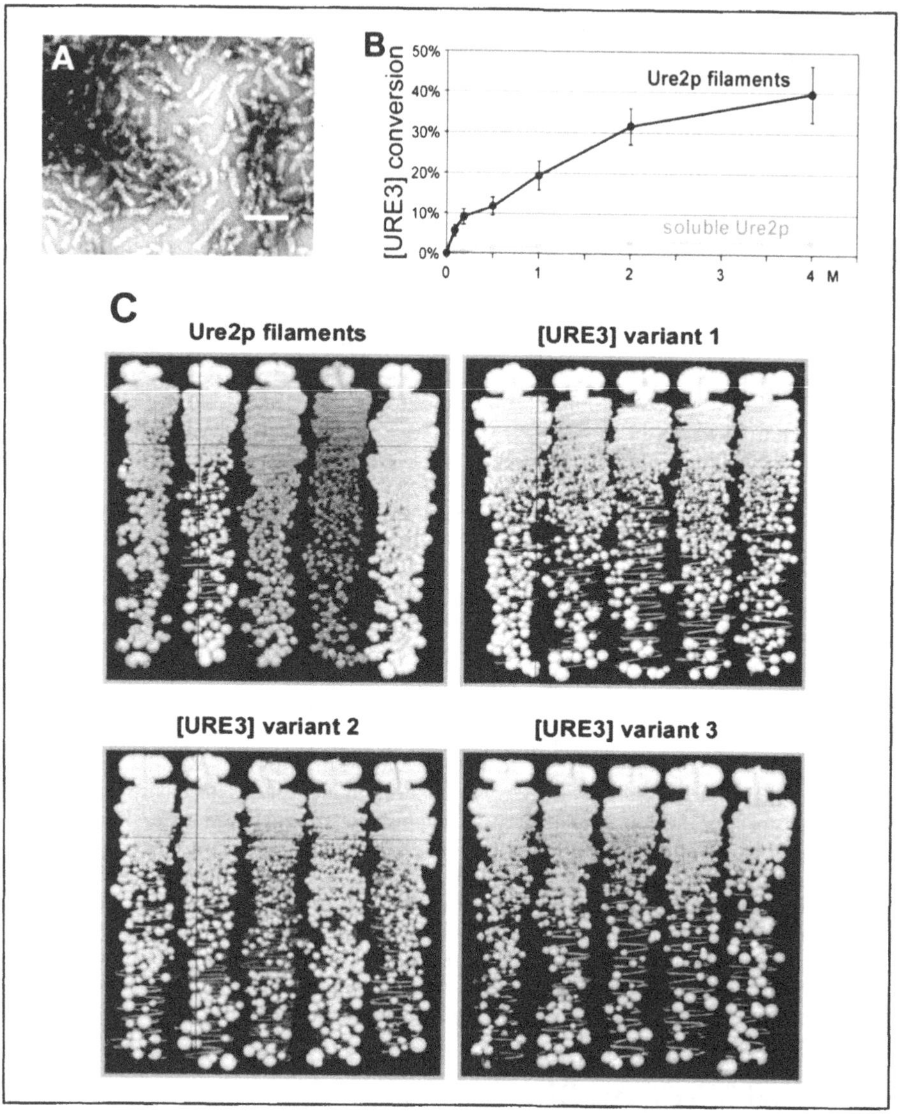

Figure 4. Amyloid of Ure2p is infectious.[56] Amyloid made in vitro from recombinant Ure2p (full length or the prion domain or fusions of the prion domain with other proteins) are infectious for yeast. A) Filaments are sonicated (bar = 100 nm) and introduced into spheroplasts with a DNA plasmid and polyethylene glycol. B) A large proportion of the clones transformed for the DNA plasmid were also infected with [URE3]. C) The infected clones included several prion variants distinguished by stability and intensity of the phenotype, here indicated by activity of a *DAL5*-promoted *ADE2* gene. Red clones have lost [URE3]. Extracts of each variant are infectious and transmit the variant of the strain from which they were made.[56]

and the L-BC virus (reviewed in ref. 62). We found that the mildly detrimental nucleic acid replicons were found in varying proportions of the wild yeast (Table 1). For example, 2 micron DNA has been shown to slow growth by 1.5-3.0%,[63] but is found in 38 of 70 wild strains.

***Table 1. Nonchromosomal genetic elements in wild* Saccharomyces**

Nonchromosomal Element	Element Present / Strains Examined
L-A dsRNA virus	15 / 70
L-BC dsRNA virus	8 / 70
20S RNA replicon	14 / 70
23S RNA replicon	1 / 70
2 μ DNA plasmid	38 / 70
[URE3] prion	0 / 70
[PSI +] prion	0 / 70
[PIN +] prion	11 / 70

Seventy wild strains of Saccharomyces, including 52 cerevisiae, 9 bayanus, 9 paradoxus isolates, were examined for the presence of the indicated RNA and DNA replicons and prions.[61]

We found that none of the wild strains carried either [URE3] or [*PSI*$^+$] (Table 1). Similarly, [*PSI*$^+$] was absent from nine clinical isolates,[64] two industrial *S. cerevisiae* and eight other non-*cerevisiae* strains of *Saccharomyces*.[65] This indicates that these prions must be quite substantially detrimental to their host. As previously reported for two clinical isolates, [*PIN*$^+$] is not rare in the wild (Table 1), but its frequency is similar to the parasitic DNA and RNA replicons, suggesting that it is a rather mild disease.

Our approach measures whether [URE3] or [*PSI*$^+$] are advantageous or not without addressing specific conditions of growth. It remains possible that there is a natural situation in which [URE3] or [*PSI*$^+$] are more of a help than a hindrance, just as the mild hemoglobin disease, Sickle Cell Trait, is an advantage in areas where malarial infection is prevalent. However, stress and the need to evolve are daily occurrences for yeast, and if [URE3] or [*PSI*$^+$] helped in this regard, they would not be hard to find in the wild.

A Self-Activating Enzyme Acting as a Prion

The word 'prion' means 'infectious protein',[66] and although most prions are found to be self-propagating amyloids, this need not be the case.[4] If an enzyme were made as an inactive precursor, and the active form of the same enzyme were necessary for activation of the precursor, then this could appear as a prion system. The vacuolar protease B of *S. cerevisiae* is made as an inactive precursor, and is normally processed proteolytically to an active form by protease A (reviewed in ref. 67). However, in mutants deleted for protease A (*pep4*Δ), evidence for some transient self-activation was obtained.[68] We showed that this self-activation of protease B could be propagated indefinitely if cells were grown on nonfermentable carbon sources, under which conditions the gene encoding protease B is derepressed.[69]

The inactive state of protease B is very stable, as is the active state. Spontaneous activation of the enzyme occurred only about once in 10^5 cells. Loss of the active state was more frequent, occuring in 1% or more of cells. The active state was transferable by cytoduction, and we called this nonchromosomal genetic element [β].[69]

[β] has all the properties expected of a prion. Growth of cells on glucose media efficiently cures [β], but from cured cells it again arises de novo (reversible curability). Overproduction of the inactive protease B precursor increases the frequency of [β] generation de novo from about 10^{-5} to about 10^{-2} or higher.[69] The propagation of [β] depends on the *PRB1* gene, but because the prion in this case is not an inactive form of the protein, the phenotype of [β] cells is the opposite of that of *prb1* mutants.

Like the [Het-s] prion of *Podospora anserina*,[70] [β] is a prion with a function for the cells. Without [β], diploid cells fail to undergo meiosis and spore formation, and die more rapidly

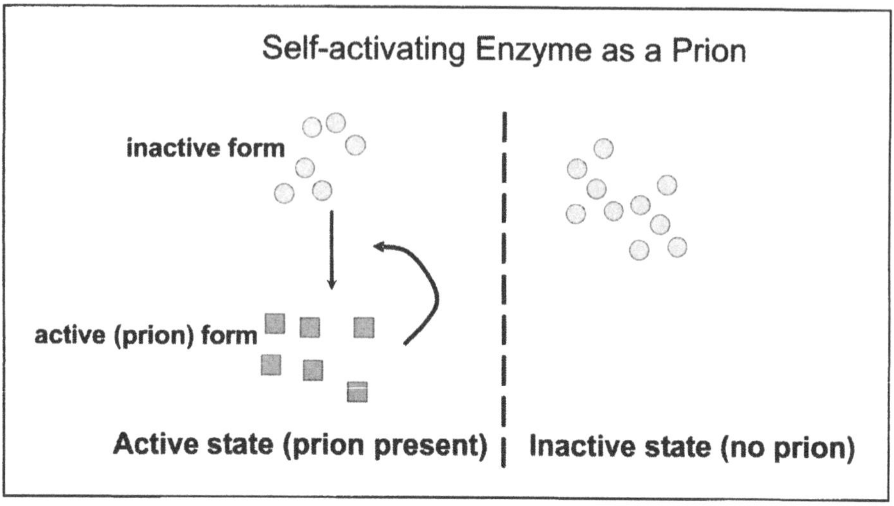

Figure 5. Enzymes needed for their own activation can be prions. "Prion" means "infectious protein", not necessarily amyloid based. If an enzyme is essential for activation of its own precursor, then cells without the active form produce the same as progeny, and those with the active form produce offspring of the same kind. Transmission of just the active form (the protein only) from one cell to another lacking it, transmits the self-propagating activity, and so is an infectious protein. Two such systems have been described, the vacuolar protease B of *S. cerevisiae*,[69] and a protein kinase of *Podospora anserina*.[72]

under starvation conditions.[69] Because [β] is only seen as a prion in the absence of protease A, one could view it as rather artificial. Alternatively, it could be seen as a prion so essential for the cell, that the protease B precursor has evolved to be protease A - cleavable, thus insuring that the prion (active protease B) is never lost. This amounts to duplication of function.

The importance of our findings is that there are many potentially self-modifying enzymes, including protein kinases, protein transacetylases, protein glycosyl transferases, protein methylases, and many others. We suggested that some of these enzymes might become prions under some circumstances. Indeed, we did not have to wait long.

A Possible Protein Kinase Prion

Crippled Growth is a nonchromosomal genetic element, called [C], of *Podospora anserina*, characterized by slow hyphal growth and dark pigmentation.[71] This trait has recently been shown to require for its propagation a gene encoding a MAP kinase kinase kinase.[72] Most strikingly, overproduction of the same enzyme increases the frequency with which the [C] nonchromosomal genetic element arises.[72] The *Crippled Growth* phenotype differs from that of mutation of the MAPKKK gene, as expected if it is due to activation of the MAPKKK enzyme, rather than inactivation. Interestingly, the MAPKKK protein has a 60 residue polyQ sequence near its N-terminus, but deletion of this sequence does not impair ability to propagate [C].[72] It is likely that [C] is a self-propagating self-activation of the MAPKKK,[73] but further work will be needed to confirm this conclusion.

Acknowledgements

This Research was supported in part by the Intramural Research Program of the NIH, NIDDK.

References

1. Prusiner SB, ed. Prion Biology and Diseases. Cold Spring Harbor: Cold Spring Harbor Laboratory Press, 2004.
2. Chesebro B. Introduction to the transmissible spongiform encephalopathies or prion diseases. Br Med Bull 2003; 66:1-20.
3. Alper T, Haig DA, Clarke MC. The exceptionally small size of the scrapie agent. Biochem Biophys Res Commun 1966; 22:278-284.
4. Griffith JS. Self-replication and scrapie. Nature 1967; 215:1043-1044.
5. Dickinson AG, Meikle VMH, Fraser H. Identification of a gene which controls the incubation period of some strains of scrapie in mice. J Comp Path 1968; 78:293-299.
6. Carlson GA, Kingsbury DT, Goodman PA et al. Linkagae of prion protein and scrapie incubation time genes. Cell 1986; 46:503-511.
7. Bolton DC, McKinley MP, Prusiner SB. Identification of a protein that purifies with the scrapie prion. Science 1982; 218:1309-1311.
8. Bueler H, Fischer M, Lang Y et al. Normal development and behavior of mice lacking the neuronal cell-surface PrP protein. Nature 1992; 356:577-582.
9. Stahl N, Borchelt DR, Hsiao K et al. Scrapie prion protein contains a phosphatidylinositol glycolipid. Cell 1987; 51:229-240.
10. Bueler H, Aguzzi A, Sailer A et al. Mice devoid of PrP are resistant to Scrapie. Cell 1993; 73:1339-1347.
11. Silveira JR, Raymond GJ, Hughson AG et al. The most infectious prion protein particles. Nature 2005; 437:257-61.
12. Legname G, Baskakov IV, Nguyen HOB et al. Synthetic mammalian prions. Science 2004; 305:673-676.
13. Collinge J. Variant Creutzfeldt-Jakob disease. Lancet 1999; 354:317-23.
14. Cooper TG. Transmitting the signal of excess nitrogen in Saccharomyces cerevisiae from the Tor proteins to th GATA factors: Connecting the dots. FEMS Microbiol Revs 2002; 26:223-238.
15. Magasanik B, Kaiser CA. Nitrogen regulation in Saccharomyces cerevisiae. Gene 2002; 290:1-18.
16. Lacroute F. Non-Mendelian mutation allowing ureidosuccinic acid uptake in yeast. J Bacteriol 1971; 106:519-522.
17. Cox BS. PSI, a cytoplasmic suppressor of super-suppressor in yeast. Heredity 1965; 20:505-521.
18. Zhouravleva G, Frolova L, LeGoff X et al. Termination of translation in eukaryotes is governed by two interacting polypeptide chain release factors, eRF1 and eRF3. EMBO J 1995; 14:4065-4072.
19. Stansfield I, Jones KM, Kushnirov VV et al. The products of the SUP45 (eRF1) and SUP35 genes interact to mediate translation termination in Saccharomyces cerevisiae. EMBO J 1995; 14:4365-4373.
20. Wickner RB. [URE3] as an altered URE2 protein: Evidence for a prion analog in S. cerevisiae. Science 1994; 264:566-569.
21. Drillien R, Lacroute F. Ureidosuccinic acid uptake in yeast and some aspects of its regulation. J Bacteriol 1972; 109:203-208.
22. Singh AC, Helms C, Sherman F. Mutation of the non-Mendelian suppressor ψ^+ in yeast by hypertonic media. Proc Natl Acad Sci USA 1979; 76:1952-1956.
23. Lund PM, Cox BS. Reversion analysis of [psi-] mutations in Saccharomyces cerevisiae. Genet Res 1981; 37:173-182.
24. Chernoff YO, Derkach IL, Inge-Vechtomov SG. Multicopy SUP35 gene induces de-novo appearance of psi-like factors in the yeast Saccharomyces cerevisiae. Curr Genet 1993; 24:268-270.
25. Masison DC, Wickner RB. Prion-inducing domain of yeast Ure2p and protease resistance of Ure2p in prion-containing cells. Science 1995; 270:93-95.
26. Masison DC, Maddelein ML, Wickner RB. The prion model for [URE3] of yeast: Spontaneous generation and requirements for propagation. Proc Natl Acad Sci USA 1997; 94:12503-12508.
27. Ter-Avanesyan A, Dagkesamanskaya AR, Kushnirov VV et al. The SUP35 omnipotent suppressor gene is involved in the maintenance of the non-Mendelian determinant [psi+] in the yeast Saccharomyces cerevisiae. Genetics 1994; 137:671-676.
28. Doel SM, McCready SJ, Nierras CR et al. The dominant PNM2⁻ mutation which eliminates the [PSI] factor of Saccharomyces cerevisiae is the result of a missense mutation in the SUP35 gene. Genetics 1994; 137:659-670.
29. DePace AH, Santoso A, Hillner P et al. A critical role for amino-terminal glutamine/asparagine repeats in the formation and propagation of a yeast prion. Cell 1998; 93:1241-1252.

30. Liu JJ, Lindquist S. Oligopeptide-repeat expansions modulate 'protein-only' inheritance in yeast. Nature 1999; 400:573-576.
31. Maddelein ML, Wickner RB. Two Prion - Inducing regions of Ure2p are nonoverlapping. Mol Cell Biol 1999; 19:4516-4524.
32. Kochneva-Pervukhova NV, Paushkin SV, Kushnirov VV et al. Mechanism of inhibition of Ψ^+ prion determinant propagation by a mutation of the N-terminus of the yeast Sup35 protein. Embo J 1998; 17:5805-10.
33. Borchsenius AS, Wegrzyn RD, Newnam GP et al. Yeast prion protein derivative defective in aggregate shearing and production of new 'seeds'. EMBO J 2001; 20:6683-6691.
34. Ross ED, Baxa U, Wickner RB. Scrambled prion domains form prions and amyloid. Mol Cell Biol 2004; 24:7206-7213.
35. Ross ED, Edskes HK, Terry MJ et al. Primary sequence independence for prion formation. Proc Natl Acad Sci USA 2005; 102:12825-12830.
36. Kochneva-Pervukhova NV, Poznyakovski AI, Smirnov VN et al. C-terminal truncation of the Sup35 protein increases the frequency of de novo generation of a prion-based [PSI$^+$] determinant in Saccharmyces cerevisiae. Curr Genet 1998; 34:146-151.
37. Pierce MM, Baxa U, Steven AC et al. Is the prion domain of soluble Ure2p unstructured? Biochemistry 2005; 44:321-8.
38. Lansbury Jr PT, Costa PR, Griffiths JM et al. Structural model for the beta-amyloid fibril based on interstrand alignment of an antiparallel-sheet comprising a C-terminal peptide. Nat Struct Biol 1995; 2:990-8.
39. Benzinger TL, Gregory DM, Burkoth TS et al. Propagating structure of Alzheimer's beta-amyloid(10-35) is parallel beta-sheet with residues in exact register. Proc Natl Acad Sci USA 1998; 95:13407-12.
40. Antzutkin ON, Balbach JJ, Leapman RD et al. Multiple quantum solid-state NMR indicates a parallel, not antiparallel, organization of beta-sheets in Alzheimer's beta-amyloid fibrils. Proc Natl Acad Sci USA 2000; 97:13045-50.
41. Petkova AT, Ishii Y, Balbach JJ et al. A structural model for Alzheimer's beta-amyloid fibrils based on experimental constraints from solid state NMR. Proc Natl Acad Sci USA 2002; 99:16742-16747.
42. Tycko R. Insights into the amyloid folding problem from solid-state NMR. Biochemistry 2003; 42:3151-3159.
43. Jayasinghe SA, Langen R. Identifying structural features of fibrillar islet amyloid polypeptide using site-directed spin labeling. J Biol Chem 2004; 279:48420-48425.
44. Der-Sarkissian A, Jao CC, Chen J et al. Structural organization of α-synuclein fibrils studied by site-directed spin labeling. J Biol Chem 2003; 278:37530-37535.
45. Yoder MD, Jurnak F. Protein motifs: 3. The parallel beta helix and other coiled folds. FASEB J 1995; 9:335-342.
46. Perutz MF, Johnson T, Suzuki M et al. Glutamine repeats as polar zippers: Their possible role in inherited neurodegenerative diseases. Proc Natl Acad Sci USA 1994; 91:5355-5358.
47. Ross ED, Minton AP, Wickner RB. Prion domains: Sequences, structures and interactions. Nat Cell Biol 2005; 7:1039-1044.
48. Taylor KL, Cheng N, Williams RW et al. Prion domain initiation of amyloid formation in vitro from native Ure2p. Science 1999; 283:1339-1343.
49. Baxa U, Taylor KL, Wall JS et al. Architecture of Ure2p prion filaments: The N-terminal domain forms a central core fiber. J Biol Chem 2003; 278:43717-43727.
50. Chan JCC, Oyler NA, Yau WM et al. Parallel β-sheets and polar zippers in amyloid fibrils formed by residues 10—39 of the yeast prion protein Ure2p. Biochemistry 2005; 44:10669-10680.
51. Baxa U, Speransky V, Steven AC et al. Mechanism of inactivation on prion conversion of the Saccharomyces cerevisiae Ure2 protein. Proc Natl Acad Sci USA 2002; 99:5253-5260.
52. Bai M, Zhou JM, Perrett S. The yeast prion protein Ure2 shows glutathione peroxidase activity in both native and fibrillar forms. J Biol Chem 2004; 279:50025-30.
53. Maddelein ML, Dos Reis S, Duvezin-Caubet S et al. Amyloid aggregates of the HET-s prion protein are infectious. Proc Natl Acad Sci USA 2002; 99:7402-7.
54. King CY, Diaz-Avalos R. Protein-only transmission of three yeast prion strains. Nature 2004; 428:319-323.
55. Tanaka M, Chien P, Naber N et al. Conformational variations in an infectious protein determine prion strain differences. Nature 2004; 428:323-328.
56. Brachmann A, Baxa U, Wickner RB. Prion generation in vitro: Amyloid of Ure2p is infectious. EMBO J 2005; 24:3082-3092.

57. Petkova AT, Leapman RD, Guo Z et al. Self-propagating, molecular-level polymorphism in Alzheimer's beta-amyloid fibrils. Science 2005; 307:262-5.
58. Eaglestone SS, Cox BS, Tuite MF. Translation termination efficiency can be regulated in Saccharomyces cerevisiae by environmental stress through a prion-mediated mechanism. EMBO J 1999; 18:1974-1981.
59. True HL, Berlin I, Lindquist SL. Epigenetic regulation of translation reveals hidden genetic variation to produce complex traits. Nature 2004; 431:184-7.
60. True HL, Lindquist SL. A yeast prion provides a mechanism for genetic variation and phenotypic diversity. Nature 2000; 407:477-483.
61. Nakayashiki T, Kurtzman CP, Edskes HK et al. Yeast prions [URE3] and [PSI$^+$] are diseases. Proc Natl Acad Sci USA 2005; 102:10575-10580.
62. Wickner RB. In: Knipe DM, Howley PM, eds. Fields Virology. Philadelphia: Lippincott, Williams and Wilkins, 2001:629-658.
63. Mead DJ, Gardner DCJ, Oliver SG. The yeast 2 μ plasmid: Strategies for the survival of a selfish DNA. Mol Gen Genet 1986; 205:417-421.
64. Resende CG, Outeiro TF, Sands L et al. Prion protein gene polymorphisms in Saccharomyces cerevisiae. Mol Microbiol 2003; 49:1005-1017.
65. Chernoff YO, Galkin AP, Lewitin E et al. Evolutionary conservation of prion-forming abilities of the yeast Sup35 protein. Mol Microbiol 2000; 35:865-876.
66. Prusiner SB. Novel proteinaceous infectious particles cause scrapie. Science 1982; 216:136-144.
67. Jones EW. Three proteolytic systems in the yeast Saccharomyces cerevisiae. J Biol Chem 1991; 266:7963-7966.
68. Zubenko GS, Park FJ, Jones EW. Genetic properties of mutations at the PEP4 locus in Saccharomyces cerevisiae. Genetics 1982; 102:679-690.
69. Roberts BT, Wickner RB. A class of prions that propagate via covalent auto-activation. Genes Dev 2003; 17:2083-2087.
70. Coustou V, Deleu C, Saupe S et al. The protein product of the het-s heterokaryon incompatibility gene of the fungus Podospora anserina behaves as a prion analog. Proc Natl Acad Sci USA 1997; 94:9773-9778.
71. Silar P, Haedens V, Rossingnol M et al. Propagation of a novel cytoplasmic, infectious and deleterious determinant is controlled by translational accuracy in Podospora anserina. Genetics 1999; 151:87-95.
72. Kicka S, Silar P. PaASK1, a mitogen-activated protein kinase kinase kinase that controls cell degeneration and cell differentiation in Podospora anserina. Genetics 2004; 166:1241-1252.
73. Wickner RB, Edskes HK, Ross ED et al. Prions of yeast are genes made of protein: Amyloids and enzymes. Cold Spring Harb Symp Quant Biol 2004; 49:489-496.

Chapter 2

The Genetic Control of the Formation and Propagation of the [*PSI*⁺] Prion of Yeast

Mick F. Tuite* and Brian S. Cox

Abstract

It is over 40 years since it was first reported that the yeast *Saccahromyces cerevisiae* contains two unusual cytoplasmic 'genetic' elements: [PSI^+] and [*URE3*]. Remarkably the underlying determinants are protein-based rather than nucleic acid-based, i.e., that they are prions, and we have already learnt much about their inheritance and phenotypic effects from the application of 'classical' genetic studies alongside the more modern molecular, cellular and biochemical approaches. Of particular value has been the exploitation of chemical mutagens and 'antagonistic' mutants which directly affect the replication and/or transmission of yeast prions. In this Chapter we describe what has emerged from the application of classical and molecular genetic studies, to the most intensively studied of the three native yeast prions, the [PSI^+] prion.

Introduction

Genetic studies with *Saccharomyces cerevisiae* first revealed the existence of non-Mendelian traits in this species over half a century ago with the discovery of the mitochondrial *petite* mutation.[1,2] A decade after the realization that the yeast cytoplasm contained such nucleic acid-based genetic determinant, two other cytoplasmic 'genetic' elements referred to as [PSI^+] and [*URE3*] emerged from yeast genetic laboratories in the UK and France respectively.[3,4] These new determinants were also originally identified by their non-Mendelian pattern of inheritance, but confounded researchers because they were not associated with mitochondrial DNA. There followed a period of considerable speculation about the molecular nature of the underlying genetic determinants[5] and it was not until some 30 years later that the 'genetic' determinants [PSI^+] and [*URE3*] were shown to be proteinaceous in nature rather than nucleic acid based i.e., that they were prions.[6] In this chapter we describe what has emerged from the application of classical and molecular genetic studies, to the most intensively studied of the yeast prions, namely the [PSI^+] prion.

A Short History of the 'ψ Factor', a Novel Non-Mendelian Element

Discovery of the 'ψ Factor'

As with many of the most important discoveries in science, the 'ψ factor' (as the [PSI^+] prion was originally known), was discovered by serendipity, emerging from a genetic analysis of mutants of *S. cerevisiae* that suppressed nonsense mutations. In these studies Brian Cox, then at the University of Oxford, was following the inheritance of *SUQ5* (also referred to as *SUP16*) a dominant nonsense suppressor and was using suppression of the nonsense

*Corresponding Author: Mick F. Tuite—Department of Biosciences, University of Kent, Canterbury, Kent CT2 7NJ, U.K. Email: m.f.tuite@kent.ac.uk.

Protein-Based Inheritance, edited by Yury O. Chernoff. ©2007 Landes Bioscience.

(UAA-ochre) *ade2-1* allele to monitor inheritance of the *SUQ5* suppressor.[3] Cells carrying the *ade2-1* allele usually form red colonies and are adenine auxotrophs (i.e., Ade⁻) but in the *SUQ5*-carrying strains being studied by Cox, this allele was suppressed by *SUQ5* thereby giving Ade⁺ cells that gave white colonies. However, in genetic crosses in which *SUQ5* was segregating, an unexpectedly high number of haploid spores gave rise to white Ade⁺ colonies with small red, Ade⁻ sectors. Subsequent genetic analysis of these 'nonsuppressed' mutants led to the discovery of the 'ψ factor'[3,7] with the nonsuppressed cells being designated ψ⁻ and the suppressed cells, ψ⁺.

The key genetic cross that resulted in Cox concluding that the 'ψ factor' was a non-Mendelian trait was one in which he crossed cells from one of the red sectors (with the genotype *SUQ5 ade2-1* ψ⁻) to a white suppressed progenitor strain (with the genotype *SUQ5 ade2-1* ψ⁺). The resulting diploid showed the suppressed phenotype i.e., ψ⁺ was dominant to ψ⁻, and all the haploid spores emerging from the diploid after meiosis showed the same nonsense suppression phenotype and a 4:0 pattern of inheritance (Fig. 1). As expected, when the *SUQ5 ade2-1* ψ⁺ strain was crossed to an *ade2-1*ψ⁻ strain lacking the *SUQ5* suppressor, each tetrad contained 2 white Ade⁺ to 2 red Ade⁻ spores confirming that the *SUQ5* mutation was still present in the strain and that the ψ⁺ mutation was not itself an efficient suppressor of *ade2-1* in the absence of *SUQ5*. Following up on his original observation, Cox went on to demonstrate that a number of different laboratory strains were ψ⁻, and that the *SUQ5* suppressor could also suppress other nonsense mutations (e.g., *his5-2, can1-100, lys1-1*), but only if the strain carried the ψ⁺ 'mutation'. Subsequently it was shown that *SUQ5* encoded a mutant form of a serine-inserting tRNA with an altered anticodon sequence such that it can decode the UAA (ochre) codon albeit inefficiently.[8] The *ade2-1* allele and the other suppressible alleles studied by Cox contained premature ochre mutations within their coding sequences. Certain nonsense alleles (e.g., *cyc1-72*) could also be suppressed by [*PSI⁺*] in the absence of a defined suppressor tRNA.[9]

The Cytoplasmic Location of the ψ Factor

While a 4:0 pattern of inheritance is entirely consistent with the ψ factor having a genetic determinant located outside the nucleus, nevertheless additional evidence to support this hypothesis was needed, especially as genetic studies by Young and Cox[10] had shown that the ψ⁺ mutation was inherited independently of mitochondrial genome markers such as erythromycin resistance (ery^r). The 'classical' method for demonstrating transmission of a cytoplasmically-located genetic determinant in fungi is cytoduction, but the formation of heterokaryons is not part of the yeast life cycle. The availability of the karyogamy-defective mutant (the *kar1* mutant[11]) which blocks the fusion of haploid nuclei from different parents during mating but has no affect on plasmogamy, provided such a genetic tool with which to confirm the cytoplasmic inheritance of the ψ factor. In a *kar1* x *KAR1⁺* cross, the parent cells fuse to form a cell containing two haploid nuclei in a mixed cytoplasm and are the equivalent of a dikaryon. New haploid daughter cells can emerge from the binucleate cell, which contain one or other of the parental nuclei, but in a cytoplasm contributed by both parents (Fig. 1). Applying this cytoduction assay to an analysis of the inheritance of the ψ factor showed that all haploid cytoductants segregating in a *kar1* x *KAR1⁺* cross were ψ⁺ irrespective of which parental nucleus was inherited from the binucleate 'heterokaryon'.[12]

With the unambiguous demonstration that the ψ factor was inherited through the cytoplasm, but was not linked genetically to the mitochondrial genome, presented a new dilemma; what was the molecular nature of the ψ factor? Several other nucleic acid species are found in the cytoplasm of *S. cerevisiae* with the linear double-stranded (ds) RNA genome of the 'killer' virus[13] and the circular ds DNA 2 μm plasmid[14] being the two major species present. Yet a variety of physical and mutagenesis studies failed to link the ψ factor with any known nucleic acid 'genome'[15] although some intriguing preliminary evidence that ψ⁻ cells could be transformed to ψ⁺ using a DNA fraction enriched for circular extrachromosomal forms of ribosomal DNA (rDNA) repeats, the so-called 3 μm circles, was subsequently reported by Dai et al.[16]

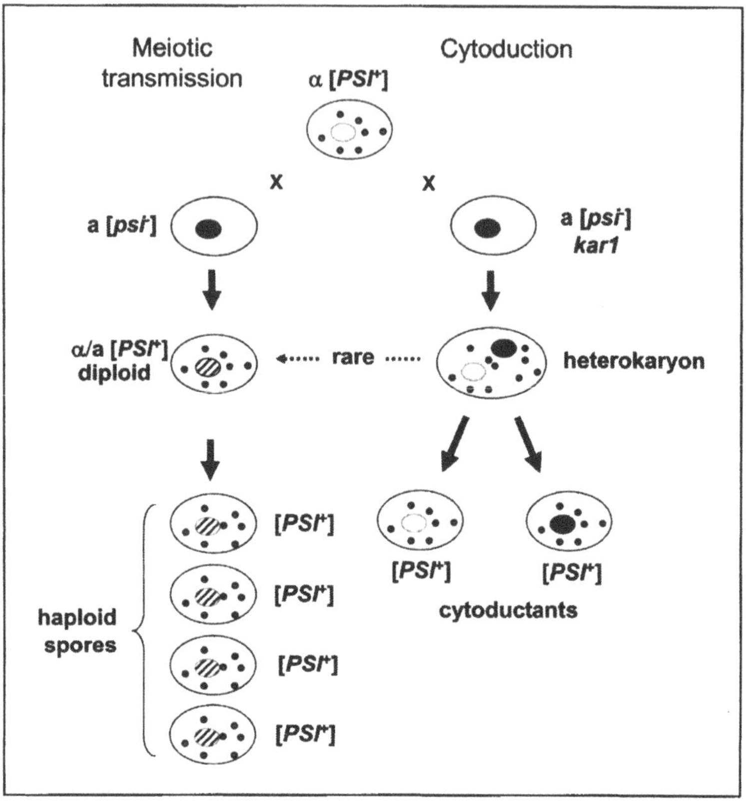

Figure 1. The yeast $[PSI^+]$ prion is transmitted as an extrachromosomal genetic element. The cytoplasm of $[PSI^+]$ cells contain a number of distinct physical entities - propagons - that are necessary for the continued propagation of the $[PSI^+]$ state. When a $[PSI^+]$ cell is mated to a propagon-free $[psi^-]$ cell (left panel) the diploid is $[PSI^+]$ and all four haploid meiotic progeny are also $[PSI^+]$. When a $[PSI^+]$ cell is mated to a $[psi^-]$ *kar1* cell that is defective in nuclear fusion (right panel) the resulting unstable heterokaryons generate new haploid cells—cytoductants. At a frequency approaching 100%, such cytoductants are $[PSI^+]$ irrespective of the haploid nucleus carried. NB: The nuclei arising from fusion between two parental nuclei are shown with hatches.

Mutagenesis Studies on the ψ Factor

Early studies on chemical or physical agents that caused a ψ^+ to ψ^- mutation proved to be informative albeit with the benefit of hindsight. On the one hand, known DNA active mutagens such as ultraviolet light (UV) and ethyl methane sulphonate (EMS) could induce this mutation but only at a frequency expected of a single nuclear gene.[5,17] These data strongly suggested that the ψ factor was dependent in some way on a single nuclear gene and that single hit mutations could result in permanent loss of the ψ factor. We now know that the gene is most likely *SUP35*.

In contrast to the findings with classical nuclear mutagens, certain non DNA active agents, in particular guanidine hydrochloride (GdnHCl) a protein denaturant, and methanol, were found to efficiently induce the ψ^+ to ψ^- 'mutation' without any apparent underlying mutagenic damage to DNA or RNA.[18] One of the possible implications of the discovery of a class of mutagens active against the ψ factor but not DNA was that the phenotype associated with the presence of the ψ factor was the result of an "epigenetic self-sustaining system" that could result in the cells adopting one of two meta-stable states i.e., ψ^+ and ψ^-.[5] The self-sustaining ψ^+ state could be perturbed either by mutation of a gene encoding a component of that system or by chemically perturbing the system in some other undefined manner.[5] As we now know, that 'state' (now more typically referred to as [*PSI⁺*]) reflects the presence of altered but self-perpetuating conformers of Sup35p, a cellular protein involved in translation termination, that is encoded by the *SUP35* gene.

[PSI⁺] *Is the Prion form of Sup35p*

In addition to proposing that the non-Mendelian inheritance of the phenotypically unrelated [*URE3*] determinant could be explained by the prion-like behaviour of the Ure2 protein (Ure2p), Wickner[6] also suggested that the [*PSI⁺*] determinant might be due to a prion. Three independent reports pointed firmly at the product of the *SUP35* gene (the Sup35 protein—Sup35p) as the most likely candidate for the prion protein associated with [*PSI⁺*]:[19] (a) the discovery that mutations in the *SUP35* gene could result in cells no longer being able to maintain the [*PSI⁺*] determinant i.e., they become [*psi⁻*],[20,21] (b) that mutations in the *SUP35* gene (originally known as *sal3* mutants[19]) gave the same phenotype as cells carrying [*PSI⁺*], and (c) that over expression of the *SUP35* gene led to the de novo appearance of [*PSI⁺*] strains.[22]

Within three years of the original Wickner paper,[6] a number of independent studies were published that provided important genetic and biochemical data that supported the hypothesis that the [*PSI⁺*] determinant (i.e., Cox's mysterious ψ factor) is the prion form of Sup35p[23-25] (recent reviews). The definitive proof of a protein-only (prion) based mechanism for [*PSI⁺*] came with the demonstration that introduction of a purified recombinant 'prion-like' form of Sup35p into a [*psi⁻*] cell by 'protein transformation' triggers a high rate of conversion of those cells to a stable [*PSI⁺*] state.[26,27]

With the recognition that [*PSI⁺*] is the prion form of Sup35p, many if not all of the unusual genetic properties that Cox described for the ψ factor could be accounted for: non-Mendelian inheritance, dependency on a nuclear gene, cytoduction and failure to associate with a known cytoplasmic nucleic acid genome (Fig. 1) although many questions still remain unanswered as will be evident from other contributions in this volume.

The [*PSI⁺*] Phenotype

As described above, the phenotype of a [*PSI⁺*] cell that is most widely exploited is that of nonsense suppression and this phenotype is a direct consequence of a defect in the translation termination machinery in [*PSI⁺*] cells. The relative efficiency with which a suppressor tRNA can translate any stop codon be it a natural terminator or a premature stop codon, is directly related to the ability of the tRNA to out compete the termination machinery in binding to the ribosomal A site where the codon is positioned (Fig. 2). Yeast, as for other eukaryotes, encodes two proteins, eRF1 and eRF3, which constitute the functional release factor that plays the central role in terminating protein synthesis in response to an in-frame stop codon.[28,29] eRF1 (encoded by the *SUP45* gene in yeast) recognises the stop codon positioned in the ribosomal A site via its N-terminal domain. The central domain of eRF1 interacts with the peptidyl-transferase centre of the ribosome to trigger the hydrolysis of the peptidyl-tRNA ester bond with the concomitant release of the completed polypeptide chain and interacts with eRF3 through its C-terminal domain.[30] eRF3 stimulates the termination reaction in a GTP-dependent manner and although the precise role of eRF3 in the termination reaction is not well defined, it may be required for the coupling of stop codon recognition by eRF1 to release of the polypeptide chain from the ribosome.[31] Deletion of either the *SUP35* or *SUP45* gene is lethal although it is conceivable that cell death is because one or both of these proteins may also play essential nontranslational roles in the cell (for an example, see ref. 32).

The strength of the nonsense suppression phenotype used to monitor the presence of the [*PSI⁺*] prion, is the net result of the interplay between three parameters (Fig. 2). Of paramount importance is the relative efficiency with which a cognate (nonsense suppressor) or

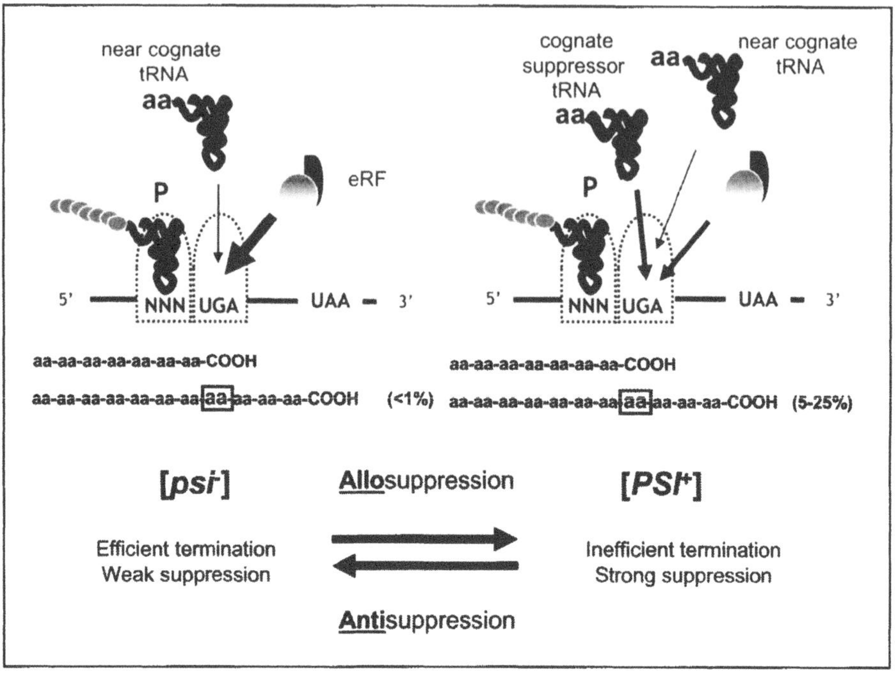

Figure 2. The molecular basis of the [PSI^+]-associated nonsense suppression phenotype. Left) When a stop codon arrives at the ribosomal A site it is efficiently recognised by the release factor which in yeast and other eukaryotes comprises of two proteins eRF1 (Sup45p) and eRF3 (Sup35p). There are some near-cognate tRNAs able to translate termination codons albeit inefficiently and these are only usually detected when the termination machinery impaired as, for example, is the case in [PSI^+] cells or in *sal* mutants. Translation then proceeds to the next in-frame stop codon, which here is UAA. Right) If the cell encodes a mutant tRNA which is able to recognise a defined stop codon i.e., a cognate nonsense suppressor tRNA such as the SUQ5-encoded $tRNA^{Ser}$ [8], this tRNA is able to compete more efficiently than a near-cognate tRNA with the release factor for the termination codon at the A site. Again the efficiency with which the cognate suppressor tRNA competes in greatly increased if termination machinery is impaired.

near cognate (wild-type) tRNA is able to compete with the translation termination machinery for the target nonsense codon. In a [PSI^+] strain, the termination machinery is defective most likely because a high proportion (>90%) of Sup35p is present in the form of prion aggregates[33,34] possibly preventing a functional interaction between Sup35p and Sup45p (eRF1). Mutations in either the *SUP35* or *SUP45* genes can also give the same termination-defective phenotype. The other two parameters that can influence suppression efficiency are the compatibility of the amino acid inserted in response to the stop codon with function of the encoded polypeptide chain, and the nucleotide sequences immediately 5' and 3' to the suppressible stop codon.[35,36]

In practical terms the most straightforward measure of termination efficiency is via a colony level analysis using strains carrying a nonsense allele of either the *ADE1* (e.g., *ade1-14*^{UGA}) or *ADE2* (e.g., *ade2-1*^{UAA}) gene. In such strains high efficiency nonsense suppression gives white Ade^+ colonies, while low efficiency nonsense suppression gives pink colonies showing a weak adenine prototrophy. Nonsuppressed strains will be red and adenine auxotrophs. A more quantitative means of determining the efficiency on nonsense suppression can be achieved through the use of plasmid-borne gene fusions in which two assayable open reading frames (ORFs) are separated by a stop codon; e.g., PGK-stop-lacZ,[37] lacZ-stop-luc.[38]

While the ability of the *SUQ5* mutation to suppress ochre alleles is absolutely dependent upon the cells being [*PSI*$^+$], this is not the case for a large number of other well characterised nonsense suppressors. For example, there are several tyrosine-inserting ochre suppressor tRNAs described e.g., *SUP4*, that can suppress *ade2-1* and other ochre alleles in a [*psi*$^-$] strain, but are lethal when crossed into a [*PSI*$^+$] genetic background.[39] This lethal phenotype presumably reflects a synergism between the suppressor tRNA and the weakened termination mechanism in the [*PSI*$^+$] background that also leads to efficient readthrough of naturally-occurring termination codons at the end of ORFs.

Nuclear Genetic Antagonists of [*PSI*$^+$]

The nonsense suppression phenotype that registers the presence of the [*PSI*$^+$] prion in a given strain provides a simple yet effective means of screening for mutations that affect the ability of a cell to propagate [*PSI*$^+$]. Thus starting with either an *ade1-14* [*PSI*$^+$] strain or an *SUQ5 ade2-1* [*PSI*$^+$] strain, mutations that result in loss of the diagnostic white Ade$^+$ phenotype can be readily detected by a visual screen for red Ade$^-$ colonies. Alternatively, positive selection of nonsuppressed mutants can be achieved by using the suppressible *can1-100* allele; [*psi*$^-$] cells expressing this allele are resistant to the toxic arginine analogue canavanine whereas in [*PSI*$^+$] cells the *can1-100* allele is suppressed and cells become canavanine sensitive. An alternative positive selectable marker for the loss of [*PSI*$^+$] has also been developed, based on *ura3-14*, a nonsense allele of the *URA3* gene.[40]

Using such phenotypic screens, two basic classes of nuclear genetic antagonists of [*PSI*$^+$] have been identified that have arisen either spontaneously or as a consequence of chemical or UV mutagenesis (Table 1):

- 'Psi-no more' (PNM) mutants which lead to a loss in the ability of a cell to propagate the [*PSI*$^+$] prion and which in turn leads to loss of the [*PSI*$^+$] phenotype; and
- 'Antisuppressor' (ASU) mutants which mask the [*PSI*$^+$] phenotype, but retain the ability to propagate the prion form of Sup35p.

For both classes of modifier either dominant (gain-of-function) or recessive (loss-of-function) mutants have been described and can be differentiated on the basis of the different patterns of inheritance shown (Fig. 3). Each class of mutant will be described in detail below.

PNM Mutants

The first 'PNM' mutant to be described was designated 'R' (red) and carried a dominant chromosomal mutation that eliminated [*PSI*$^+$] from cells.[41] When the R mutant was crossed to a [*PSI*$^+$] strain not only was the resulting diploid [*psi*$^-$], but so were all four spores (Fig. 3). However, for a given tetrad only two of the spores retained the dominant PNM character whereas the other two spores gave rise to cells with the normal genetic behaviour of a [*psi*$^-$]

Table 1. Nuclear genetic antagonists of the [PSI$^+$]-associated nonsense suppression phenotype

Gene	Gene Product/Function
Antisuppressors (ASU)	
ASU1-8	tRNA modification
MOD5	Delta 2-isopentenyl pyrophosphate:tRNA isopentenyl transferase, tRNA modification
PSI-No-More (PNM)	
PNM1	Hsp104, a molecular chaperone implicated in protein disaggregation
PNM2	Allele of *SAL3* (*SUP35*)

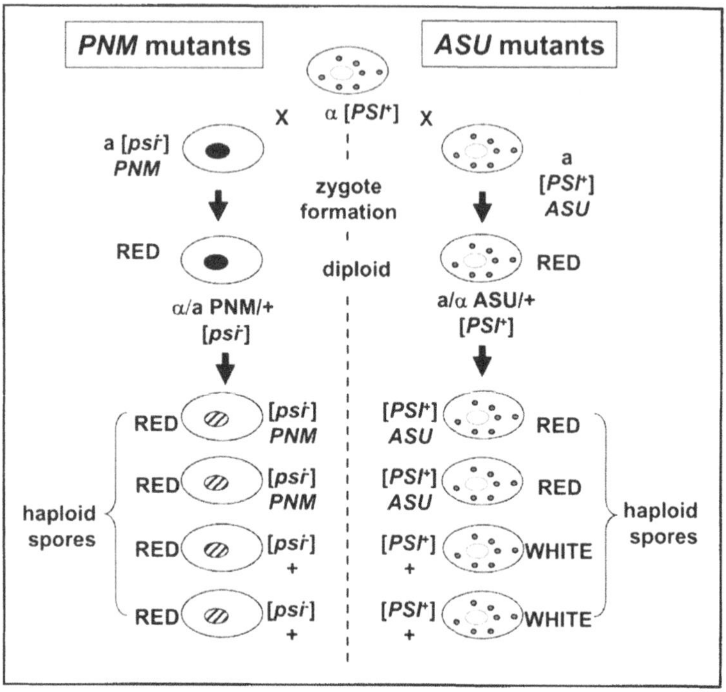

Figure 3. A genetic cross can differentiate between mutations that eliminate the [PSI^+] prion (*PNM*, 'PSI-No-More' mutants) and dominant antisuppressor mutants (*ASU*). The diploid formed between a *PNM* mutant and a [PSI^+] strain loses the ability to replicate the [PSI^+] prion because it can no longer efficiently produce the new propagons required. Consequently all of the haploid meiotic spores do not inherit the ability to propagate the [PSI^+] state i.e., are [psi^-]. In contrast, in a cross between an *ASU* mutant and a [PSI^+] strain, although the phenotype is the same as the *PNM/+* diploid, the *ASU/+* diploid still has the ability to generate new propagons. The reason the nonsense suppression phenotype is not expressed is because these cells also produce a significant level of soluble and functional Sup35p which ensures that efficient translation termination occurs. Consequently those meiotic spores that inherit the *ASU* mutation show a red nonsuppressed phenotype whereas those that do not, show the [PSI^+] phenotype.

strain i.e., were no longer 'PNM'. The underlying genetic defect in the R mutant was therefore nuclear in nature but resulted in loss of the extrachromosomal [PSI^+] determinant. The loss of the [PSI^+] determinant was not immediate because if a newly formed [PSI^+] x [psi^-] PNM zygote was abruptly induced to go into meiosis i.e., without further rounds of cell division, a significant number of [PSI^+] spores can be found and the longer the diploid is grown before meiosis is induced, the fewer the number of [PSI^+] spores are observed until most if not all spores give rise to [psi^-] cells.[7,42]

To date only two *PNM* genes (*PNM1* and *PNM2*) have been identified via the genetic analysis of dominant 'PNM' alleles. A comprehensive screen for nonsuppressing revertants of a [PSI^+] strain[12] has also identified a large number of recessive *pnm* mutants although the number of complementation groups these define is not known. In contrast to a dominant *PNM* mutant, a recessive *pnm* mutant yields a suppressed [PSI^+] diploid when crossed with a [PSI^+] strain and the resulting diploid yields suppressed ([PSI^+]) and nonsuppressed ([psi^-]) spores in various ratios but most typically 2 [PSI^+]:2 *pnm* [psi^-].

The *PNM1* Gene

The *PNM1* gene is allelic to the *HSP104* gene[43] a finding that is in keeping with the earlier discovery that the product of this gene, the molecular chaperone Hsp104, is essential for the maintenance of the [*PSI*$^+$] prion.[44] Hsp104 is an ATPase that works in combination with the Hsp70 and Hsp40 chaperone to remodel proteins that have aggregated in stressed cells. Hsp104 therefore plays an important role in helping cells recover from temperature stress, but also functions at much lower temperatures to facilitate the propagation of all three yeast prions.[45,46]

A number of mutations in the *HSP104* gene have been described which result in a defect in [*PSI*$^+$] propagation and these mutations are usually located in one of the two domains implicated in ATP binding and/or hydrolysis (Fig. 4) which is to be expected given that this activity is crucial to Hsp104's role as a protein disaggregase. Some of the *PNM* alleles of *HSP104* so far described, including a gene disruption, are recessive with respect to the [*PSI*$^+$] propagation defect. However other alleles, for example the original *PNM1-1* allele described by Young & Cox[41] and a double mutation that inactivates both nucleotide binding domains (the K218T/K620T),[44] both show a dominant phenotype with respect to [*PSI*$^+$] loss. Sequence analysis of the *PNM1-1*

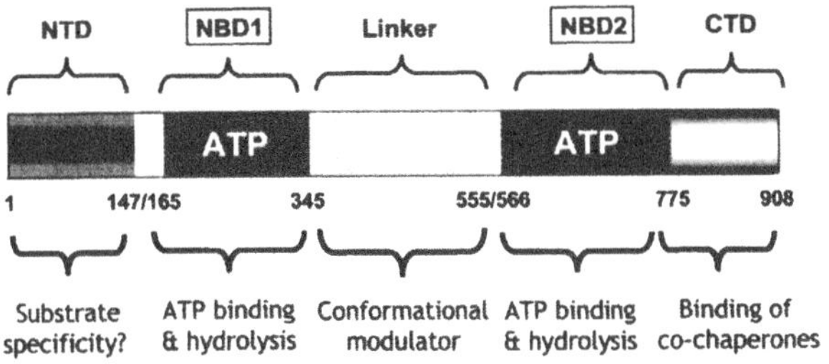

Mutation	Hsp104 Defect	Dominant/Recessive	Ref
D184Y	NBD1?	ND	66
K218T	NBD1	Recessive[1]	44
T317A	NBD1, sensor 1	Recessive	81
K620T	NBD2	Recessive[1]	44
E645T	NBD2	ND	47,81
Y662A/K	substrate threading	ND	47,81
F722S	NBD2	ND	80
N728A	NBD2, sensor 1	Recessive	80
R826M	sensor 2	Recessive	83

Figure 4. Mutations within the molecular chaperone Hsp104 can give rise to a 'PSI-No-More' phenotype. Upper: A schematic representation of the key functional domains of Hsp104 showing the two nucleotide binding domains (ATP; NBD1/NBD2) and the N-terminal (NTD) and C-terminal (CTD) domains plus the linker region between the two NBDs. The proposed functional roles of the different domains are indicated below. See references 45 and 46 for further discussion on the functional organisation of Hsp104. Lower: mutations that have so far been described which give a 'PNM' phenotype. The location of the mutations is given together with an indication of whether the mutation in question is dominant or a recessive with respect to the 'PNM' phenotype. ND indicates that the dominance/recessive character of the mutant was not reported. [1] The single mutations K218T and K620T are also dominant/semi-dominant PNMs in some genetic backgrounds.

allele has revealed mutations in both nucleotide binding domains NBD1 and NBD2.[43] A dominant PNM phenotype is probably due to a defect in Hsp104 oligomerisation and/or the generation of nonfunctional Hsp104 hexamers containing both mutant and wild-type Hsp104.

Recent studies have shown that the N-terminal region of Hsp104 (residues 1-147) is dispensable not only for [*PSI*$^+$] propagation but also protein refolding and thermotolerance[47] indicating that this region of the protein is not required to carry out these functions. In contrast, any mutation in the *HSP104* gene that leads to a PNM phenotype impairs the disaggregase function of the chaperone and is consistent with current models for the role of Hsp104 in prion propagation, namely that the chaperone breaks down prion aggregates to generate new prion seeds necessary for continued propagation of the prion state[45,46] (see also chapter by Cox et al).

The *PNM2* Gene

The *PNM2* gene is allelic to *SUP35*[21] and the discovery that a mutation in the *SUP35* gene could affect the maintenance of the [*PSI*$^+$] determinant provided a crucial piece of evidence that linked Sup35p with the [*PSI*$^+$] prion.[20,21,23] The *PNM2-1* allele sequenced by Doel et al[21] contains a Gly to Asp substitution at residue 58 near the amino terminus of the protein in a region shown by deletion studies to be required for [*PSI*$^+$] propagation, i.e., in the prion-forming domain (PrD)[20,48] (Fig. 5). The mutation was in the second of five oligopeptide repeats located between residues 41 and 97 of Sup35p; the so-called oligopeptide repeat (OPR) region. This region of the Sup35p protein together with the amino terminal Gln/Asn-rich 40 residues (also known as the QN-rich - QNR - region) constitute the prion-forming domain that is essential for the aggregation (the QNR region) and the continued propagation (the OPR region) of the [*PSI*$^+$] prion.[49] Why the mutant Sup35p encoded by the *PNM2-1* allele (i.e., Sup35p$^{PNM2\text{-}1}$) has a dominant negative property with respect to prion propagation remains to be established, but it has been noted that the degree of dominance shown by this allele with respect to its 'PNM' phenotype is sensitive to the genetic background of the strain or the [*PSI*$^+$] variant present in the strain in which it is introduced.[50]

As for the wild-type Sup35p, the Sup35p$^{PNM2\text{-}1}$ protein is able to induce [*PSI*$^+$] de novo when overexpressed in a [*PIN*$^+$][*psi*] strain[50,51] and can also form protein aggregates in vivo and self-seeding amyloid-like aggregates in vitro.[49,51] This shows that the mutant Sup35p$^{PNM2\text{-}1}$ protein is able to take up its prion form but once established that form is not efficiently propagated. This in turn leads to a mitotic instability and loss of [*PSI*$^+$]. In a *PNM2-1*/+ heterozygote presumably mixed prion aggregates are formed which are impaired in their role in propagation of the prion state.

In addition to the original *PNM2-1* allele, several other *PNM* alleles of *SUP35* have been generated by random or site-directed mutagenesis.[52,53] Among the mutants described by De Pace et al[52] were *PNM* alleles which contained single amino acid substitutions located between residues 9 and 33 in the QNR region of the Sup35p-PrD important for protein aggregation, rather than in the OPR region where the *PNM2-1* mutation lies. In the QNR region mutants, the soluble mutant Sup35p$^{PNM2\text{-}1}$, molecules are poorly recruited into the prion-like aggregates of Sup35p but, importantly also prevent the generation of new prion seeds required for continued propagation of the [*PSI*$^+$] prion.[52] The *PNM* mutants reported by King[53] were single amino acid substitutions within the Sup35p-QNR region, but these mutations had differing PNM properties depending on the [*PSI*$^+$] variant present in the strain (see below).

Surprisingly, even though single amino acid substitutions in either the Sup35p-QNR or Sup35p-OPR regions can lead to a defect in the propagation of the [*PSI*$^+$] prion, the primary amino acid sequence of the QNR+OPR regions of Sup35p per se, does not seem to be critical for prion formation. This has recently emerged from a study in which the amino acid sequence of Sup35p between residues 3 and 114 were 'randomized' and the new 'scrambled' sequence used to replace the wild type sequence. That these engineered Sup35p 'mutants' were still mostly able to form and propagate the [*PSI*$^+$] prion suggest that it is the length and/or amino

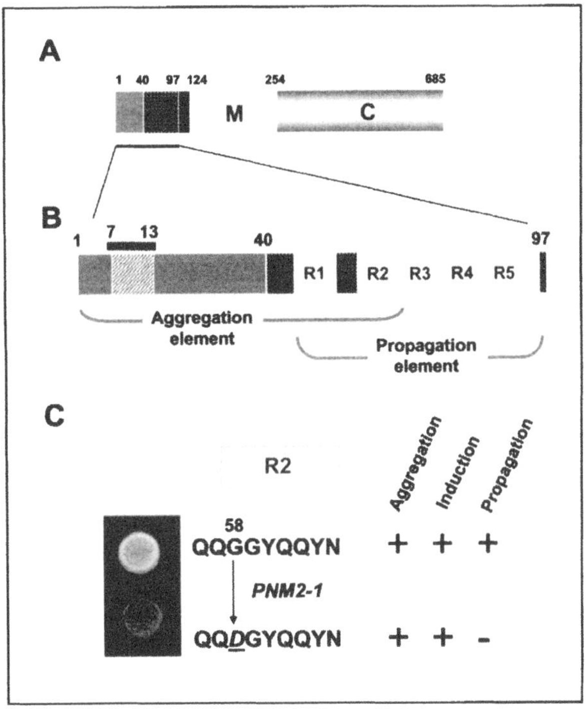

Figure 5. The prion-forming domain of the Sup35p protein. A) The 685 residue Sup35p protein has three regions defined by the locations of the first three in-frame AUG (Met) codons: the amino terminal N domain (residues 1-123) that is absolutely required for the prion behaviour of the protein; the highly charged middle region (M; residues 124-253) and the carboxyl-terminal domain (C; residues 254-685) which carries the essential release factor activity.[20,48] B) Within the N domain lies a region, the so-called prion-forming domain—which contains a region important for aggregation and a separate but overlapping region important for propagation of the prion state.[48] The aggregation element contains a short peptide sequence based around residues 7 to 13, that is highly amyloidogenic. The propagation element contains five imperfect copies of an oligopeptide repeat sequence (R1-R5). C) The 'PSI-No-More' mutant *PNM2-1* carries a single amino acid substitution (Gly to Asp) within repeat R2 of the propagation element. This single substitution leads to a form of Sup35p that can still aggregate but can no longer be efficiently propagated in most laboratory strains.[20,48,50]

acid composition of the Sup35p-PrD that is critical for the prion-like behaviour of Sup35p, a conclusion that emerged from parallel studies with the Ure2p prion protein.[54] The apparent contradiction between these two sets of observations remains to be resolved. Certainly large scale substitutions or deletions within the adjacent OPR region can generate 'PNM' alleles of *SUP35*.[20,55,56] What the findings of Ross et al[54] may indicate is that when Sup35p (and Ure2p) form amyloid-like fibres, the interacting prion protein molecules may take up a parallel in-register β-sheet structure rather than an anti-parallel structure.[57]

Antisuppressor (ASU) Mutants

While studies on the PNM mutants have made important contributions to our understanding of what makes a Sup35p a prion protein and the role of Hsp104 in propagation of the prion state, much less work has been done on antisuppressor mutants which prevent the [*PSI⁺*] phenotype from being expressed but without impairing propagation of the [*PSI⁺*] prion (Table

1). Such mutants are distinct from the recessive antisuppressor mutants originally described by McCready and Cox[58] since this latter class (which map to at least 8 different loci, designated *ASU1 - ASU8*) are most likely mutations that affect the structure and/or function of the suppressor tRNA. For example, one of the *ASU* genes (also called *MOD5*) is required for the synthesis of N^6-delta 2-(isopentenyl) adenosine, a modified tRNA nucleoside important for codon-anticodon recognition.[59] Another class of yeast antisuppressor described by Chernoff and colleagues[60] were single nucleotide changes in the 18S ribosomal RNA that reduced the efficiency of the nonsense suppression associated with certain *sup35* and *sup45* alleles that give rise to an omnipotent suppression phenotype that results in suppression and hence readthrough of all three stop codons. The molecular basis for the recessive antisuppression phenotype seen in both the tRNA modification and 18S rRNA mutants is most likely that they reduce the efficiency with which a suppressor tRNA or a naturally-occurring suppressor tRNA competes with the release factor-mediated termination event for the premature nonsense codon (see Fig. 2). There is no evidence that these recessive antisuppressors affect the propagation of the [*PSI*$^+$] prion.

Single amino acid substitutions within the Sup35p-PrD[52] or deletion of the Sup35p-PrD[20,48,55] can lead to a dominant antisuppressor phenotype in a [*PSI*$^+$] cell. This is because these mutations generate forms of Sup35p that remain largely soluble because the efficiency with which they are seeded by the endogenous [*PSI*$^+$] Sup35p seeds in vivo and their ability to form amyloid fibrils in vitro are dramatically reduced.[52] The soluble Sup35p^{ASU} molecules still contain the functional C-terminal region of the Sup35p molecule required for translation termination and their presence in a cell, irrespective of whether or not the cell is [*PSI*$^+$], results in a shift in the balance towards termination and against nonsense suppression i.e., antisuppression. This contrasts to Sup35p$^{PNM2\text{-}1}$ molecules which can interact with the endogenous [*PSI*$^+$] seeds but, in so doing, block the ability to generate the new seeds required for continued propagation.[49,51,52]

Guanidine Hydrochloride-Induced [psi$^-$] Mutants

Even before the [*PSI*$^+$] determinant had been identified as a 'protein-only' element, a number of chemical agents that were not mutagenic for DNA or RNA-based determinants, had been identified that resulted in efficient elimination of the [*PSI*$^+$] determinant from growing cells. These agents include methanol, dimethyl sulphoxide,[18] high osmolarity,[61] the kastellpaolitines[62] and the actin cytoskeleton disruptor, latrunculin A.[63]

By far the most effective [*PSI*$^+$] 'curing agent' described, and the one we best understand in terms of mode of action, is the protein denaturant guanidine hydrochloride (GdnHCl). When GdnHCl is added to growing [*PSI*$^+$] cells this results in an essentially prion-free [*psi*$^-$] culture after 10 to 12 generations of growth with the [*psi*$^-$] cells appearing after only 4 or 5 generations of growth.[18,64] The emergence of [*psi*$^-$] cells appears to be a consequence of the GdnHCl, inhibiting even at low concentrations (1-5 mM), the key ATPase activity of the molecular chaperone Hsp104.[65-67] This in turn results in an inability of the [*PSI*$^+$] cell to generate the new prion seeds required for continued propagation of the [*PSI*$^+$] prion. Consequently, the seeds present in the [*PSI*$^+$] cell at the time the GdnHCl is added, are diluted by cell division and eventually seed-free and hence prion-free [*psi*$^-$] cells emerge. For a detailed discussion of this mechanism and how it can be exploited to gain insights into the nature of the [*PSI*$^+$] seeds see chapter by Cox et al.

[*psi*$^-$] mutants generated by treatment of [*PSI*$^+$] cells with these chemically diverse collection of compounds do not undergo any permanent change in their nuclear genotype that blocks prion propagation. This can be shown by the reintroduction of prion seeds back into the [*psi*$^-$] mutant by either genetic back crossing or cytoduction (see Fig. 1) which in both cases reestablishes a stable [*PSI*$^+$] cell. While the [*psi*$^-$] mutants induced by most of these agents are able to revert spontaneously to [*PSI*$^+$] at a frequency of approximately 10^{-5} to 10^{-6}, those induced by GdnHCl treatment do not revert spontaneously back to [*PSI*$^+$] at any detectable frequency.[68,69]

Although at the time this observation was made it was thought that this was due to a physical deletion of the [*PSI⁺*] 'determinant', we now know that the reason is that the GdnHCl treatment also eliminates all other prions from the yeast cell including the [*PIN⁺*] prion that facilitates the de novo formation of [*PSI⁺*].[70,71] That spontaneous reversion to [*PSI⁺*] is seen with other chemically-induced [*psi⁻*] mutants would suggest that these 'mutagens' act on a different target that only affects the [*PSI⁺*] prion and do not generate [*pin⁻*] cells.

[*PSI⁺*] 'Variants': Mutants without Genetic Change

One of the more remarkable properties of prions both in yeast and in animals is that they can exist in different conformational states that modify the associated neuropathology of the disease (for animal PrP) or the associated phenotype (as is the case for all three yeast prions).[72] In yeast such prion strains are referred to as 'variants' in order not to confuse with yeast 'strains' that may have different phenotypes due to underlying differences in genotypes: yeast prion variants show different phenotypes but the amino acid sequence of the prion protein is identical in the different variants.

Yeast prion variants were first described for [*PSI⁺*] by the Liebman group when they noted that [*PSI⁺*] strains generated de novo in the same experiment often had different yet stably inherited phenotypes as defined using suppression of the *ade1-14* marker i.e., colony color and degree of adenine prototrophy.[70,71] Two basic [*PSI⁺*] variants have been described and are usually referred to as 'weak' and 'strong' reflecting directly the efficiency of nonsense suppression: weak variants show low efficiency of suppression (i.e., strong translation termination) while strong variants show efficient suppression.[70,71,73] Prion variants also show other readily scorable differences; for example weak variants show reduced mitotic stability and an elevated amount of soluble Sup35p compared with a strong variant.[73]

The first clue to the physical basis for distinct yeast prion variants came from in vitro seeding studies which showed that the Sup35p aggregates found in a weak [*PSI⁺*] variant are less efficient at seeding soluble Sup35p than aggregates from a strong [*PSI⁺*] variant.[73] This difference in the efficiency of Sup35p conversion would therefore lead to weak variants having higher levels of soluble Sup35p, which is the case.[73] The higher proportion of soluble—and therefore functional—Sup35p in the weak variants would give the characteristic antisuppressor, low efficiency suppression phenotype (see Fig. 2 and previous section). Direct conformation that the different [*PSI⁺*] variants result from distinct yet heritable conformers of Sup35p came from two groups using novel 'protein transformation' assays.[26,27] Tanaka et al[26] generated in vitro, distinct amyloid-like forms of a recombinant fragment of Sup35p derived from the N-terminus (Sup35NMp; Fig. 5), by using different temperatures for the polymerisation. When these different forms of Sup35NMp were introduced into [*psi⁻*] cells, they gave rise to distinct [*PSI⁺*] variants that were stably propagated over many subsequent generations of growth. For example, Sup35NMp aggregates formed at 4°C gave rise to strong variants whereas aggregates that formed at 37°C gave rise to weak variants. Therefore the Sup35p protein can take up at least three different self-replicating conformational states each of which results in different levels of soluble Sup35p in the cell and hence different [*PSI⁺*]-associated phenotypes.

An elegant model to explain the physical basis of prion variants in yeast has recently been presented by Weissman and colleagues.[74] In this model - which they validate experimentally - different variants are generated as a consequence of the dynamic interplay between several different parameters including conformation-dependent differences in the rate prion aggregates are formed in the cell and the rate of fragmentation. The Sup35p aggregates in a strong variant grow more slowly than the aggregates found in the weak variants, but the key difference is the fragility of the aggregates formed; in the strong variant these aggregates are much more susceptible to breakage than the faster forming aggregates in the weak variants and so the number of prion seeds was greater in the strong variants.[74] It remains to be seen whether this model can also explain different mammalian prion strains.

While most studies on [PSI^+] variants compare weak vs strong, several other [PSI^+] variants have also been described. For example King[53] described three different [PSI^+] variants that showed distinct phenotypes when introduced into a series of different yeast strains carrying defined *sup35* mutations. As with the weak and strong variants, protein aggregates from strains carrying these [PSI^+] variants (called [VH], [VK] and [VL]) could be used to infect [psi^-] cells and the resulting [PSI^+] cells had the phenotype associated with the original variant. Sup35p fibrils formed by these variants also show distinct conformational differences.[75]

The existence of stable [PSI^+] prion variants which have identical nuclear genotypes but which show significant differences in the efficiency with which the translation termination machinery can recognise a stop codon amply illustrates the epigenetic nature of the yeast [PSI^+] prion and also the potential and varied impact the prion can have on cell phenotype.

Translation Termination in [PSI^+] Strains

The [PSI^+] prion provides the yeast cell with a novel epigenetic mechanism with which to modulate translation termination thereby facilitating the translation of in-frame stop codons. This of course raises two important evolutionary questions: "Why would such a mechanism have evolved?" and "What evolutionary benefit—if any—does it confer?" These questions are addressed in the chapters by Wickner et al and Zhouravleva et al.

As is now well established, [PSI^+] strains show a measurable defect in translation termination as defined by their ability to suppress a range on nonsense alleles i.e., in [PSI^+] strains the ribosome is able to translate stop codons, albeit inefficiently (see above). Yet it must be remembered that all 6000+ ORFs in the yeast genome are terminated by a stop codon. Any efficient translation of a natural terminator would result in the addition of C-terminal sequences to the encoded polypeptide chain. This in turn could have an adverse effect on function, turnover and/or localisation of that polypeptide. This presumption is well supported by several observations; for example, the reported lethal interactions between certain efficient suppressor tRNAs and [PSI^+][39] (see above) and between *sal3* alleles of *SUP35* and [PSI^+].[19]

Can translation readthrough of a natural terminator actually occur in a [PSI^+] strain? Analysis of natural terminators at the 3' ends of validated yeast ORFs has indicated that there is an over representation of in-frame stop codons 3' to the authentic terminator. Such 'tandem stops' should reduce any potentially detrimental effects of efficient readthrough.[76,77] This is not the case for all natural terminators however; for example the yeast genome contains a number of instances of two ORFs separated by a stop codon and in 8/58 cases reported by Namy et al[38,78] there was a high level of readthrough of the terminator separating the two ORFs (efficiency between 3 and 25%). Intriguingly only two of these eight showed a significant reduction in readthrough in a [psi^-] strain suggesting that there might also be [PSI^+]-independent modulation of termination in yeast. There has also been a report that readthrough of the UAG terminator of the *S. cerevisiae PDE2* gene that encodes a high affinity cAMP phosphodiesterase generates a modified form of the protein that is C-terminally extended by 20 residues and is unstable compared to the 'normal' Pde2p product.[38] In a [PSI^+] strain translation readthrough of the 'natural' *PDE2* stop codon was elevated some 20-fold compared to a [psi^-] strain and the consequence of readthrough was an increase in the cellular concentration of cAMP.[38] [PSI^+] may have its phenotypic effects via changing readthrough of only a small number of 'natural' terminators in yeast mRNAs. A further possibility is that the [PSI^+]-induced phenotypic variation may also reflect changes in the rate of decay of mRNAs whose natural terminators are being readthrough, so-called 'nonstop mRNA decay'.[79]

The very existence of the [PSI^+] prion in laboratory strains of *S. cerevisiae* is therefore something of an evolutionary puzzle, but nevertheless provides researchers with a powerful tool with which to rapidly explore some of the key questions in prion biology: what makes a prion, how is the prion state propagated and how does a cell react to the presence of a prion? Discovering why it exists may be somewhat more of a challenge.

Acknowledgements

Research on the yeast [*PSI*+] prion, carried out in the authors' laboratory, is funded by the Biotechnology and Biological Sciences Research Council (BBSRC), the Wellcome Trust, the EC (through the APOPIS project: LSHM-CT-2003-503330) and a personal award to BSC by the Leverhulme Trust.

References

1. Ephrussi B. Nucleo-cytoplasmic relations in micro-organisms. Oxford University Press, 1953.
2. Sherman F. Respiration-deficient mutants of yeast. I. Genetics. Genetics 1963; 48:375-385.
3. Cox B. PSI, a cytoplasmic suppressor of super-suppressor in yeast. Heredity 1965; 20:505-521.
4. Lacroute F. Non-Mendelian mutation allowing ureidosuccinic acid uptake in yeast. J Bact 1971; 106:519-522.
5. Cox BS, Tuite MF, McLaughlin CS. The ψ factor of yeast: A problem in inheritance. Yeast 1988; 4:159-178.
6. Wickner RB. [URE3] as an altered URE2 protein: Evidence for a prion analog in Saccharomyces cerevisiae. Science 1994; 264:566-569.
7. Cox BS. Psi Phenomena in Yeast. In: Hall MD, Linder P, eds. Early Days of Yeast Genetics. NY: Cold Spring Harbor Laboratory Press, 1993:219-239.
8. Waldron C, Cox BS, Wills N et al. Yeast ochre suppressor SUQ5-ol is an altered tRNA Ser UCA. Nucleic Acids Res 1981; 9:3077-3088.
9. Liebman SW, Sherman F. Extrachromosomal psi+ determinant suppresses nonsense mutations in yeast. J Bacteriol 1979; 139:1068-1071.
10. Young CS, Cox BS. Extrachromosomal elements in a super-suppression system of yeast. II. Relations with other extrachromosomal elements. Heredity 1972; 28:189-199.
11. Conde J, Fink GR. A mutant of Saccharomyces cerevisiae defective for nuclear fusion. Proc Natl Acad Sci USA 1976; 73:3651-3655.
12. Cox BS, Tuite MF, Mundy CR. Reversion from suppression to nonsuppression in SUQ5 [psi+] strains of yeast: The classification of mutations. Genetics 1980; 95:589-609.
13. Wickner RB. Double-stranded and single-stranded RNA viruses of Saccharomyces cerevisiae. Annu Rev Microbiol 1992; 46:347-75.
14. Velmurugan S, Mehta S, Uzri D et al. Stable propagation of 'selfish' genetic elements. J Biosci 2003; 28:623-636.
15. Tuite MF, Lund PM, Futcher AB et al. Relationship of the [psi] factor with other plasmids of Saccharomyces cerevisiae. Plasmid 1982; 8:103-111.
16. Dai H, Tsay SH, Lund PM et al. Transformation of psi- Saccharomyces cerevisiae to psi+ with DNA copurified with 3 micron circles. Curr Genet 1986; 11:79-82.
17. Tuite MF, Cox BS. Ultraviolet mutagenesis studies of [psi], a cytoplasmic determinant of Saccharomyces cerevisiae. Genetics 1980; 95:611-630.
18. Tuite MF, Mundy CJ, Cox BS. Agents that cause a high frequency of genetic change from [psi+] to [psi-] in Saccharomyces cerevisiae. Genetics 1981; 98:691-711.
19. Cox BS. Allosuppressors in yeast. Genet Res 1977; 30:187-205.
20. Ter-Avanesyan MD, Dagkesamanskaya AR, Kushnirov VV et al. The SUP35 omnipotent suppressor gene is involved in the maintenance of the non-Mendelian determinant [psi+] in the yeast Saccharomyces cerevisiae. Genetics 1994; 137:671-676.
21. Doel SM, McCready SJ, Nierras CR et al. The dominant PNM2- mutation which eliminates the psi factor of Saccharomyces cerevisiae is the result of a missense mutation in the SUP35 gene. Genetics 1994; 137:659-670.
22. Chernoff YO, Derkach IL, Inge-Vechtomov SG. Multicopy Sup35 gene induces de novo appearance of Psi-like factors in the yeast Saccharomyces cerevisiae. Curr Genet 1993; 24:268-270.
23. Cox BS. Prion-like factors in yeast. Curr Biol 1994; 4:744-748.
24. Tuite MF, Cox BS. Propagation of yeast prions. Nat Rev Mol Cell Biol 2003; 4:878-890.
25. Shorter J, Lindquist SL. Prions as adaptive conduits of memory and inheritance. Nat Rev Genet 2005; 6:435-450.
26. Tanaka M, Chien P, Naber N et al. Conformational variations in an infectious protein determine prion strain differences. Nature 2004; 428:265-267.
27. King CY, Diaz-Avalos R. Protein-only transmission of three yeast prion strains. Nature 2004; 428:319-323.
28. Stansfield I, Jones KM, Kushnirov VV et al. The products of the SUP45(eRF1) and SUP35 genes interact to mediate translation termination in Saccharomyces cerevisiae. EMBO J 1995; 14:4365-4373.

29. Zhouravleva G, Frolova L, Le Goff X et al. Termination of translation in eukaryotes is governed by two interacting polypeptide chain release factors, eRF1 and eRF3. EMBO J 1995; 14:4065-4072.
30. Song H, Mugnier P, Das AK et al. The crystal structure of human eukaryotic release factor eRF1 - Mechanism of stop codon recognition and peptidyl-tRNA hydrolysis. Cell 2000; 100:311-321.
31. Salas-Marco J, Bedwell DM. GTP hydrolysis by eRF3 facilitates stop codon decoding during eukaryotic translation termination. Mol Cell Biol 2004; 24:7769-7778.
32. Valouev IA, Kushnirov VV, Ter-Avanesyan MD. Yeast polypeptide chain release factors eRF1 and eRF3 are involved in cytoskeleton organization and cell cycle regulation. Cell Motil Cytoskel 2002; 52:161-173.
33. Paushkin SV, Kushnirov VV, Smirnov VN et al. Propagation of the yeast prion-like [psi+] determinant is mediated by oligomerization of the SUP35-encoded polypeptide chain release factor. EMBO J 1996; 15:3127-3134.
34. Patino MM, Liu JJ, Glover JR et al. Support for the prion hypothesis for inheritance of a phenotypic trait in yeast. Science 1996; 273:622-626.
35. Bonetti B, Fu L, Moon J et al. The efficiency of translation termination is determined by a synergistic interplay between upstream and downstream sequences in Saccharomyces cerevisiae. J Mol Biol 1995; 251:334-345.
36. Mottagui-Tabar S, Tuite MF, Isaksson LA. The influence of 5' codon context on translation termination in S. cerevisiae. Eur J Biochem 1998; 257:249-254.
37. Firoozan M, Grant CM, Duarte J et al. Quantitation of readthrough of termination codons in yeast using a novel gene fusion assay. Yeast 1991; 7:173-183.
38. Namy O, Duchateau-Nguyen G, Hatin I et al. Identification of stop codon readthrough genes in Saccharomyces cerevisiae. Nucleic Acids Res 2003; 31:2289-2296.
39. Cox BS. A recessive lethal super-suppressor mutation in yeast and other psi phenomena. Heredity 1971; 26:211-232.
40. Manogaran AL, Kirkland KT, Liebman SW. An engineered nonsense URA3 allele provides a versatile system to detect the presence, absence and appearance of the [PSI$^+$] prion in Saccharomyces cerevisiae. Yeast 2006; 23:141-147.
41. Young CSH, Cox BS. Extrachromosomal elements in a super-suppression system of yeast. I. A nuclear gene controlling the inheritance of the extrachromosomal elements. Heredity 1971; 26:413-422.
42. McCready SJ, Cox BS, McLaughlin CS. The extrachromosomal control of nonsense suppression in yeast: An analysis of the elimination of [psi+] in the presence of a nuclear gene PNM. Mol Gen Genet 1977; 150:265-270.
43. Cox BS, Jones KM, Ho, HL et al. Manuscript in preparation.
44. Chernoff YO, Lindquist SL, Ono B. et al. Role of the chaperone protein Hsp104 in propagation of the yeast prion-like factor [psi$^+$]. Science 1995; 268:880-884.
45. True HL. The battle of the fold: Chaperones take on prions. Trends Genet. 2006; 22:110-117.
46. Bosl B, Grimminger V, Walter S. The molecular chaperone Hsp104-A molecular machine for protein disaggregation. J Struct Biol 2006; 156:139-148.
47. Hung GC, Masison DC. N-terminal domain of yeast hsp104 chaperone is dispensable for thermotolerance and prion propagation but necessary for curing prions by hsp104 overexpression. Genetics 2006; 173:611-620.
48. Ter-Avanesyan MD, Kushnirov VV, Dagkesamanskaya AR et al. Deletion analysis of the SUP35 gene of the yeast Saccharomyces cerevisiae reveals two nonoverlapping functional regions in the encoded protein. Mol Microbiol 1993; 7:683-692.
49. Osherovich LZ, Cox BS, Tuite MF et al. Dissection and design of yeast prions. PLoS Biology 2004; 2:442-451.
50. Derkatch IL, Bradley ME, Zhou P et al. The PNM2 mutation in the prion protein domain of SUP35 has distinct effects on different variants of the [PSI$^+$] prion in yeast. Curr Genet 1999; 35:59-67.
51. Kochneva-Pervukhova NV, Paushkin SV, Kushnirov VV et al. Mechanism of inhibition of Psi$^+$ prion determinant propagation by a mutation of the N-terminus of the yeast Sup35 protein. EMBO J 1998; 17:5805-5810.
52. De Pace AH, Santoso A, Hillner P et al. A critical role for amino-terminal glutamine/asparagine repeats in the formation and propagation of a yeast prion. Cell 1998; 93:1241-1252.
53. King CY. Supporting the structural basis of prion strains: Induction and identification of [PSI] variants. J Mol Biol 2001; 307:1247-1260.
54. Ross ED, Edskes HK, Terry MJ et al. Primary sequence independence for prion formation. Proc Natl Acad Sci USA 2005; 102:12825-12830.

55. Liu JJ, Lindquist S. Oligopeptide-repeat expansions modulate 'protein-only' inheritance in yeast. Nature 1999; 400:573-576.
56. Parham SN, Resende CG, Tuite MF. Oligopeptide repeats in the yeast protein Sup35p stabilize intermolecular prion interactions. EMBO J 2001; 20:2111-2119.
57. Ross ED, Minton A, Wickner RB. Prion domains: Sequences, structures and interactions. Nat Cell Biol 2005; 7:1039-1044.
58. McCready SJ, Cox BS. Antisuppressors in yeast. Mol Gen Genet 1973; 124:305-320.
59. Laten H, Gorman J, Bock RM. Isopentenyladenosine deficient tRNA from an antisuppressor mutant of Saccharomyces cerevisiae. Nucleic Acids Res 1978; 5:4329-4342.
60. Chernoff YO, Newnam GP, Liebman SW. The translational function of nucleotide C1054 in the small subunit rRNA is conserved throughout evolution: Genetic evidence in yeast. Proc Natl Acad Sci USA 1996; 93:2517-2522.
61. Singh A, Helms C, Sherman F. Mutation of the non-Mendelian suppressor, Psi, in yeast by hypertonic media. Proc Natl Acad Sci USA 1979; 76:1952-1956.
62. Bach S, Talarek N, Andrieu T et al. Isolation of drugs active against mammalian prions using a yeast-based screening assay. Nat Biotechnol 2003; 21:1075-1081.
63. Bailleul-Winslett PA, Newnam GP, Wegrzyn RD et al. An antiprion effect of the anticytoskeletal drug latrunculin A in yeast. Gene Expr 2000; 9:145-156.
64. Eaglestone S, Ruddock LW, Cox BS et al. Guanidine hydrochloride blocks a critical step in the propagation of the prion-like determinant [PSI⁺] of Saccharomyces cerevisiae. Proc Natl Acad Sci USA 2000; 97:240-244.
65. Ferreira PC, Ness F, Edwards SR at al. The elimination of the yeast [PSI⁺] prion by guanidine hydrochloride is the result of Hsp104 inactivation. Mol Microbiol 2001; 40:1357-1369.
66. Jung GM, Jones G, Masison DC. Amino acid residue 184 of yeast Hsp104 chaperone is critical for prion-curing by guanidine, prion propagation, and thermotolerance. Proc Natl Acad Sci USA 2002; 99:9936-9941.
67. Grimminger V, Richter K, Imhof A et al. The prion curing agent guanidinium chloride specifically inhibits ATP hydrolysis by Hsp104. J Biol Chem 2004; 279:7378-7383.
68. Lund PM, Cox BS. Reversion analysis of [psi⁻] mutants in Saccharomyces cerevisiae. Genet Res 1981; 37:173-182.
69. Koloteva-Levin N, Merritt GH, Tuite MF. Manuscript in preparation.
70. Derkatch IL, Chernoff YO, Kushnirov VV et al. Genesis and variability of [PSI] prion factors in Saccharomyces cerevisiae. Genetics 1996; 144:1375-1386.
71. Derkatch IL, Bradley ME, Zhou P et al. Genetic and environmental factors affecting the de novo appearance of the [PSI⁺] prion in Saccharomyces cerevisiae. Genetics 1997; 147:507-519.
72. Bessen RA, Marsh RF. Distinct PrP properties suggest the molecular basis of strain variation in transmissible mink encephalopathy. J Virol 1994; 68:7859-7868.
73. Uptain SM, Sawicki GJ, Caughey B et al. Strains of [PSI⁺] are distinguished by their efficiencies of prion-mediated conformational conversion. EMBO J 2001; 20:6236-6245.
74. Tanaka M, Collins SR, Toyama BH et al. The physical basis of how prion conformations determine strain phenotypes. Nature 2006, (in press).
75. Diaz-Avalos R, King CY, Wall J et al. Strain-specific morphologies of yeast prion amyloid fibrils. Proc Natl Acad Sci USA 2005; 102:10165-10170.
76. Williams I, Richardson J, Starkey A et al. Genome-wide prediction of stop codon readthrough during translation in the yeast Saccharomyces cerevisiae. Nucleic Acids Res 2004; 32:6605-6616.
77. Liang H, Cavalcanti AR, Landweber LF. Conservation of tandem stop codons in yeasts. Genome Biol 2005; 6:R31.
78. Namy O, Duchateau-Nguyen G, Rousset JP. Translational readthrough of the PDE2 stop codon modulates cAMP levels in Saccharomyces cerevisiae. Mol Microbiol 2002; 43:641-652.
79. Wilson MA, Meaux S, Parker R et al. Genetic interactions between [PSI⁺] and nonstop mRNA decay affect phenotypic variation. Proc Natl Acad Sci USA 2005; 102:10244-10249.
80. Hattendorf DA, Lindquist SL. Cooperative kinetics of both Hsp104 ATPase domains and interdomain communication revealed by AAA sensor-1 mutants. EMBO J 2002; 21:12-21.
81. Lum R, Tkach JM, Vierling E et al. Evidence for an unfolding/threading mechanism for protein disaggregation by Saccharomyces cerevisiae Hsp104. J Biol Chem 2004; 279:29139-29146.
82. Tkach JM, Glover JR. Amino acid substitutions in the C-terminal AAA+ module of Hsp104 prevent substrate recognition by disrupting oligomerization and cause high temperature inactivation. J Biol Chem 2004; 279:35692-35701.
83. Hattendorf DA, Lindquist SL. Analysis of the AAA sensor-2 motif in the C-terminal ATPase domain of Hsp104 with a site-specific fluorescent probe of nucleotide binding. Proc Natl Acad Sci USA 2002; 99:2732-2737.

CHAPTER 3

A Short History of Small *s*:

A Prion of the Fungus *Podospora anserina*

Sven J. Saupe*

Abstract

Prions are infectious proteins. In fungi, prions correspond to non-Mendelian genetic elements whose mode of inheritance has long eluded explanation. The [Het-s] cytoplasmic genetic element of the filamentous fungus *Podospora anserina*, was originally identified in 1952 and recognized as a prion nearly half a century later. The present chapter will attempt to describe the work on [Het-s] from a historical perspective. The initial characterisation and early genetic and physiological studies of [Het-s] are described together with the isolation of [Het-s] encoding gene. More recent work that led to the construction of a structural model for this prion is also discussed.

The Discovery of "petit *s*"

[Het-s] is a prion of the filamentous fungus *Podospora anserina*. [Het-s] was first identified as a non-Mendelian genetic element in 1952 by George Rizet, the founder of the french fungal genetics school. He discovered [Het-s] during the genetic dissection of a phenomenon typical of filamentous fungi and termed heterokaryon incompatibility.[1] Filamentous fungi grow as an interconnected network of filaments forming a syncitial structure (Fig. 1). Fusion of somatic cells readily occurs between strains, leading to the spontaneous formation of vegetative heterokaryons.[2] Many if not all organisms forming somatic chimeras posses genetic systems that allow them to distinguish somatic self from nonself. Fungi are no exception in that regard.[3] It is believed that the role of these systems is to prevent different forms of somatic cell parasitism. Viability and fitness of fungal heterokaryons is thus genetically controlled by a set of genes termed *het* genes.[4] *Het* genes exist as two or more polymorphic allelic variants and only heterokaryons in which both nuclear components have compatible *het* gene constitution are viable. A genetic difference at any of 9 *het* loci found in Podospora leads to an incompatibility reaction resulting in severe growth inhibition and cell death of the heterokaryotic cells. Heterokaryon incompatibility systems also act as barriers limiting the transmission of deleterious cytoplasmic replicons such as mycoviruses or senescence plasmids. Transmission of such replicons between incompatible strains is reduced or abolished. In Podospora, the incompatibility reaction can easily be detected at the macroscopic level by confronting strains of solid medium. An incompatible interaction results in the formation of an abnormal contact line termed barrage. Thus wild-type strains can easily be classified into so-called vegetative compatibility groups (VCG), strains that are compatible are grouped within the same VCG.

The story of [Het-s] begins when Georges Rizet observed formation of a barrage between strains that where originally compatible (i.e., belonging to the same compatibility group).[1]

*Sven J. Saupe—Laboratoire de Génétique Moléculaire des Champignons UMR 5095 CNRS Université de Bordeaux 2, 1 rue Camille St Saëns, 33077 Bordeaux cedex, France. Email: sven.saupe@ibgc.u-bordeaux2.fr

Protein-Based Inheritance, edited by Yury O. Chernoff. ©2007 Landes Bioscience.

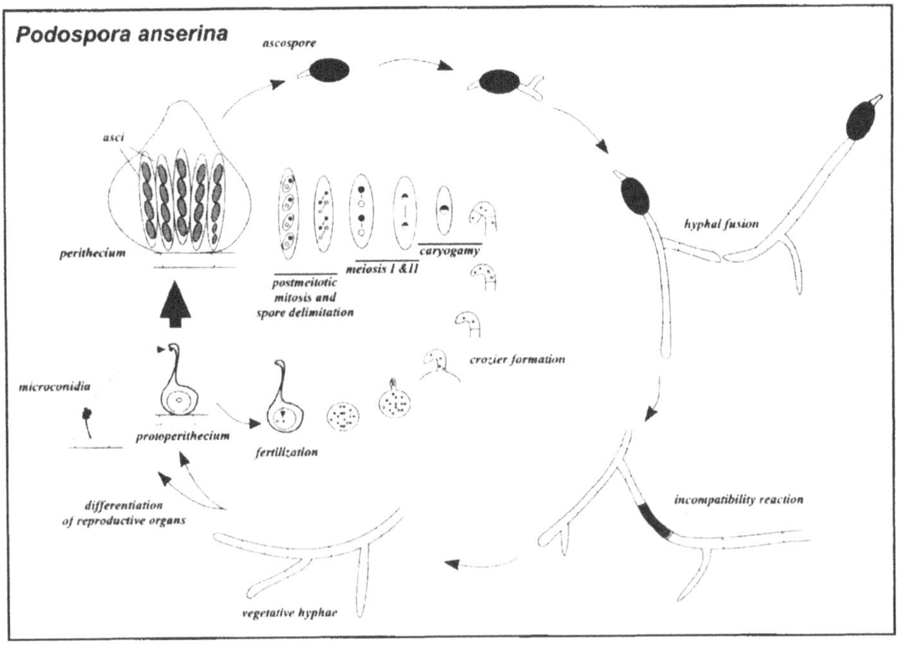

Figure 1. Life cycle of *Podospora anserina. Podospora anserina* is a coprophilic filamentous ascomycetes. The ascospores are the meiotic progeny and constitute the resistance form. After ascospore germination the fungus grows as a network of interconnected vegetative filaments -hyphae- with incomplete crosswalls. These hyphae can spontaneously fuse. If hyphal fusion involves incompatible strains the mixed fusion cell undergoes cell death. Upon nutrient starvation and exposure to light, the mycelium differentiates both male and female reproductive organs. The male gametes are termed microconidia. The female gamete is a large cell termed ascogonium and is contained in an organ called protoperithecium. After fertilization the male nucleus reaches the ascogonium and male and female nuclei undergo several mitoses. At this stage there is a transition from a coenocytic to a cellular state. A pair of nuclei of opposite mating-type enter a specialized hypha -the ascogenous hypha- and form a structure termed crozier, an isolated binucleate cell is formed in which meiosis takes place. This cycle is repeated over and over again so that a single fertilization event actually leads to about 50 independent meioses. After a post-meiotic mitosis, nuclei are packed into ascospores, each containing two nonbrother nucleic of the same half tetrad. As a result, ascospores will be homokaryotic for markers showing first division segregation but heterokaryotic for markers showing second division segregation (as in the example depicted here). The mating-type locus shows over 98% second division segregation, ascospores thus nearly invariably contain a nucleus of the + mating-type and a nucleus of the - mating type and thus give rise to self-fertile strains. The *het-s* locus is closely linked to the centromere and thus shows 95% first division segregation, thus in a *het-s* x *het-S* cross, asci nearly invariably contain two *het-s* and two *het-S* spores. Occasional mispackage events lead to formation of mononucleate ascospores and thus give rise to 5 spored asci. Strains originating from such mononucleate ascospores are single mating-type and thus self-sterile. They are used for genetic analyses. Note that the organism is haploid throughout its life cycle, the only diploid stage is the zygote.

Rizet designated these strains *s* and *S* (Table 1). To try to understand the event that led to emergence of this new incompatibility, he crossed *s* and *S* and noted a very unorthodox segregation of the parental phenotypes. Indeed, he recovered an expected 50% *S* progeny but no *s* progeny (Fig. 2). Instead the remaining strains displayed a novel phenotype that he designated s^S; s^S strains are compatible both with *s* and *S*. This nomenclature was meant to stress that the

Table 1. [Het-s] nomenclature

het-s	the *het-s* allele and the *het-s* locus
het-S	the *het-S* allele
HET-s	the *het-s*-encoded protein
HET-S	the *het-S*-encoded protein
[Het-s*]	the prion-free phenotype of a *het-s* strain
[Het-s]	the prion infected phenotype of a *het-s* strain and the prion *per se*
[Het-S]	the phenotype of *het-S* strains
s	original designation of the [Het-s] phenotype, incompatible with *S*
s^S	original designation of the [Het-s*] phenotype, compatible with *s* and *S*
S	original designation of the [Het-S] phenotype, incompatible with *s*

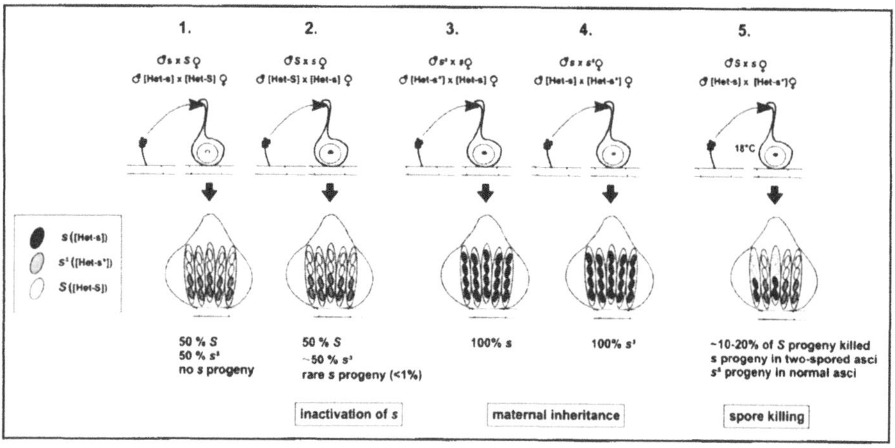

Figure 2. Properties of the *s* element ([Het-s]) during the sexual cycle. *s* and *S* strains are incompatible at the vegetative stage. In a *s* x *S* sexual cross (1 and 2), one recovers the expected 50% *S* progeny but no *s* progeny are recovered instead a novel phenotype designated *sS* appears, *sS* strains are compatible both with *s* and *S*. Even when the *s* parent is used as maternal parent -contributing most of the cytoplasm to the zygote- there are very few *s* progeny; Janine Beisson reports that in the analysis of 727 asci, only one contained *s* progeny. In s x *sS* crosses (3 and 4), all progeny have the phenotype of the maternal parent, which illustrates maternal inheritance of the *s* element. In a s x S cross performed at low temperature (18°C) when *s* is used as maternal parent (5), there is a specific abortion of *S* spores (spore killing), the *s* gene exerts a meiotic drive effect. In two-spored asci, the two remaining spores display the *s* phenotype. Color coding is black for *s*, grey for *sS* and white for *S*. (after Rizet, 1952, Beisson-Schecroun 1962, Bernet 1965).

s character had been modified by its interaction with *S*. He then observed a phenomenon that he called "réversion" by which s^S strains spontaneously reacquire the *s* phenotype. This reversion, he found, is an invasive phenomenon that rapidly spreads to the whole thallus and is transmitted to other s^S strains by simple contact. He also described that the *s* and s^S characters display maternal inheritance so that in a *s* x s^S cross the progeny will uniformly display the phenotype of the maternal parent, the one that is contributing the cytoplasm to the zygote (Fig. 2) Rizet concluded from these studies that *s* is a cytoplasmic heritable particle that can be inactivated by the *S* factor. His exact words are that *s* strains contain a particle that is absent or exists in a modified form in s^S strains.

Janine Beisson, as a student of Rizet, then examined the characteristic of the s^S to *s* reversion. She showed that the reversion process requires cytoplasmic contact as occurring during hyphal fusion.[5] She measured the rate of the reversion process (the rate of propagation of the *s* element in a s^S mycelium) and showed that this process is about ten times faster than the radial growth rate of the fungus (that is up to 70 mm per day). Janine Beisson was also able to show that the *s* element can be lost vegetatively without the intervention of the *S* factor when mycelium is regenerated from cellular structures containing very little cytoplasm. Léon Belcour later showed that the *s* element can also be lost during generation of isolated cells by protoplast formation, meaning that upon regeneration of protoplasts a small fraction of the regenerating mycelia become s^S.[6] He made the reasonable inference that "The simplest interpretation of these results is that passive and random distribution of cytoplasm occurs during protoplast formation. Those protoplasts receiving the *s* cytoplasmic factor would yield *s* mycelia. Those not receiving it would yield s^S mycelia". On the basis of this line of reasoning, he estimated that there are 1.8 to 4.6 *s* units per protoplast or 7 to 17 *s* units per 1000 μ^3 which is about similar to the number of nuclei.

By then the genetics of the *s* element were exquisitely described but naturally the physical basis of this non-Mendelian genetic element remained totally elusive. Still the "particular" nature of *s* had already been well recognized in those clever and careful genetic and physiological approaches. Janine Beisson proposed in 1962 that the *s* element regulates its own synthesis. Such a positive feed-back loop could explain the cytoplasmic transmission and maternal inheritance of the *s* trait.

Molecular Cloning of *het-s* and *het-S*, Some Answers, More Puzzles

In the late 1980s the development of a DNA transformation procedure for Podospora permitted the molecular characterisation of *s* and *S* genes. Joel Bégueret (another former student of Rizet) undertook the molecular cloning of *s* making use of an inactive allele of the *s* locus termed s^X identified in a natural isolate of *P. anserina*. Strains carrying the s^X allele are compatible both with *s* and *S* strains and cannot be converted to the *s* phenotype after contact with a wild-type *s* strain.[7,8] Soon after its rebirth in the molecular age, *s* was rebaptised to conform to the nomenclature of the other fungal heterokaryon incompatibility genes. The *s* gene became *het-s* while the *s* and s^S phenotypes became [Het-s] and [Het-s*] respectively (Table 1).

The *het-s* gene was cloned using a functional complementation approach. A *het-*s^X strain was transformed with a *het-s* genomic DNA library and transformants were scored for appearance of the [Het-s] phenotype.[7] This approach allowed the cloning of the *het-s* and *het-S* alleles. Both alleles were found to encode proteins of 289 amino acids in length differing by 13 amino acid residues. *het-s* and *het-S* knock-out strains are both viable and fully fertile.[8] The *het-*s^X allele derives from the *het-s* allele and contains a 46 bp duplication in the open reading frame leading to a premature stop codon at position 165.[9] The existence of a natural isolate in which *het-s* is in a pseudogene state might suggest that the *het-s* activity is not critical for survival even in the wild.

In order to identify which of the 13 polymorphic positions are responsible for the *het-s* and *het-S* allele specificities, Carole Deleu carried out an extensive mutational analysis.[9] She thus found that a single amino acid substitution is sufficient to turn a HET-S protein into a protein of the HET-s specificity. Amino acid 33 is a histidine in HET-S and a proline in HET-s. A H33P point mutation turns the HET-S into a protein of HET-s specificity. In other words a strain expressing the HET-S 33P mutant is no longer incompatible with [Het-s] but instead becomes incompatible with [Het-S]. This means that in this particular case incompatibility is triggered by interaction of two proteins that differ by a single amino acid (!). In addition, as wild-type HET-s, this mutant strain can display the alternate [Het-s*] and [Het-s] states. Many other mutations including deletion of H33 turn HET-S to the HET-s specificity. The reciprocal specificity switch requires a double mutation in positions 23 and 33. In other words HET-s D23A P33H behaves as HET-S. During the course of this work, Carole Deleu also characterized a remarkable allele of *het-s*, *het-s 33H*. Strains carrying this allele express the [Het-S] phenotype but can acquire the [Het-s] state after contact with a wild-type [Het-s] strain. Existence of this schizophrenic allele shows that the [Het-s] and [Het-S] states can be expressed alternatively from the same allele. So in addition to the [Het-s*] to [Het-s] transition there is the possibility of a second epigenetic switch from [Het-S] to [Het-s]. One wonders how these mutant strains manage the crisis situation that occurs during this switch as this transition necessarily leads to a transitory self-incompatibility.

The next puzzle brought forth by the molecular cloning of *het-s* and *het-S* was that of the evolutionary history of these alleles. Because Rizet saw the sudden emergence of the *s* and *S* characters, he was convinced that one allele derived from the other by a spontaneous mutation that occurred in the lab. The molecular cloning of the alleles clearly contradicted this view; *het-s* and *het-S* alleles are too divergent for that. In addition, divergence between *het-s* and *het-S* is not limited to base substitutions in the ORFs. The *het-s* allele (but not *het-S*) contains the scar of a transposable element called REPA in its promoter region and in addition the 3' region of the *het-s* locus contains a region of DNA of several kilobases that is totally absent from the genome of the *S* strain.[7-10] Several french wild-type isolates of *P. anserina* have been analysed at the *het-s* locus and both allelic types were found to be well represented. The striking observation was that all *het-s* alleles are identical to the base pair.[9] The same is true for all *het-S* alleles. It appears that two discrete highly divergent alleles are fixed in the population with no frequent intermediate type. We have yet to come up with a convincing evolutionary scenario that could account for these observations. What is clear is that polymorphism at *het-s* is ancient.

Identification of [Het-s] as a Prion Form of the HET-s Protein

Once antibodies to HET-s were developed it was determined that the HET-s protein is present both in [Het-s*] and [Het-s] strains. In fact the protein is more abundant in [Het-s*] than in [Het-s] strains;[11] electrophoretic mobility of HET-s is identical in [Het-s] and [Het-s*]. At that point, a positive feed-back loop model on the protein synthesis per se had to be rejected. Alternatively one could envision that HET-s was capable of self-activation at the post-translational level and that this post-translational modification would not alter electrophoretic mobility in SDS-PAGE.

The prion hypothesis for [Het-s] is considered by Carole Deleu in her thesis manuscript defended in 1993 and it was also suggested in a letter send to Joël Bégueret by a perspicacious reader of a 1993 paper on *het-s* mutants. But the development of the prion hypothesis for [Het-s] had to await the remarkable demonstration of the existence of yeast prions by Reed Wickner.[12] It then became evident that the properties of the [Het-s] system could be explained in a model stating that the [Het-s] non-Mendelian genetic element corresponds to the prion form of the HET-s protein. It was shown that over-expression of the *het-s* gene increases spontaneous appearance of [Het-s] and that HET-s can adopt a protease resistant state in [Het-s] strains. Interestingly, Joël Bégueret showed that conversion of a mycelium from the [Het-s*] to the [Het-s] state can occur in the presence of cycloheximide and thus does apparently not require de novo protein synthesis.[11] This observation echoes the recent elegant demonstration that conversion of the Sup35 protein to the prion state can occur from mature protein.[13]

The [Het-s] prion however displays an important genetic difference when compared to the prototypal yeast [URE3] and [*PSI*$^+$] prions. [URE3] and [*PSI*$^+$] are detected as total or partial loss of function of Ure2p and Sup35p proteins respectively, thus presence of the prion form leads to a similar phenotype as genetic inactivation of the corresponding gene. In the case of [Het-s], the prion form is not detected as a loss of function of the corresponding protein. Rather transition to the prion state is detected because the protein gains a novel property (that of triggering cell death when interacting with HET-S). In that the [Het-s] system is more similar to mammalian prions or also to the [*PIN*$^+$] yeast prion.[14]

HET-s Aggregation and Generation of [Het-s]-Infectivity in Vitro

Using GFP fusion proteins and fractionation techniques it could be shown that the HET-s protein aggregates specifically upon transition to the prion state.[15] The protein exists as a soluble monomer in [Het-s*] strains and forms high molecular weight aggregates in [Het-s] strains when highly expressed. Renatured purified recombinant histidine tagged HET-s protein is initially soluble and monomeric and spontaneously undergoes a conformational transition to an aggregated fibrillar form.[16] HET-s aggregates are typical amyloids showing a high β-sheet content, proteinase K resistance, and Congo red birefringence.[16]

The prion hypothesis for [Het-s] could be directly proven by introducing amyloid aggegates of HET-s into a [Het-s*] mycelium.[17] This was achieved using a biolistic approach. A [Het-s*] mycelium overlaid with HET-s protein was bombarded with tungsten particles to force the protein into the cells. This procedure induced appearance of the [Het-s] prion state at an elevated frequency (up to 99% efficiency). Neither the soluble form of the protein nor amorphous heat denatured HET-s aggregates induced appearance of [Het-s]. This experiment provided a direct proof of the protein-only hypothesis in the case of [Het-s] and also strongly suggested that amyloid aggregates of HET-s constitute the infectious prion species. It should however be noted that the fact that amyloids of HET-s are infectious does not prove that the entity propagating in vivo is an amyloid form of HET-s.

Structural Characterisation of HET-s

Studies of the [URE3] and [*PSI*$^+$] yeast prions revealed the existence of so-called prion forming domains (PFD). Prion forming domains are discrete regions of the prion protein both necessary and sufficient for conferring prion behaviour. In the case of yeast prions, these prion forming domains are N-terminal and characterized by a strong bias in amino acid composition as they are very rich in asparagine and glutamine residues.[18]

When submitted to proteinase K digestion, amyloid fibrils of HET-s display a 7-8 kDa resistant fragment that could be identified as the C-terminal part of the protein ranging from residue 218 to 289.[19] Proteinase K treated fibrils retain their fibrillar state (although the fibril width decreases) and retain infectivity.[17] These observations suggested that this C-terminal region could constitute the PFD of HET-s. In vitro and in vivo studies confirmed that the 218-289 region is both necessary and sufficient for [Het-s] prion activity.[19,20] It could be shown that soluble HET-s displays two distinct domains, a N-terminal α-helical globular domain spanning residue 1 to about 220-240 and a C-terminal highly flexible tail. It is this flexible tail that undergoes a major conformational transition from a random coil to a β-sheet rich structure upon aggregation[19,21] (Fig. 3).

The sequences appended to the prion forming domain strongly influence the type of supramolecular assemblies formed in vivo. In particular, it was found that in constrast to HET-s-GFP and HET-s(218-289)-GFP fusion proteins which form dot-like aggregates in vivo, the HET-s(157-289)-GFP fusion protein (which retains only a small part of the globular domain region) forms elongated aggregates in vivo that can reach up to 150 μ in length.[22]

A combination of hydrogen/deuterium exchange, solid state NMR and mutational approaches established a structure model for HET-s(218-289) in its amyloid conformation.[23] NMR approaches delimited four β-strands (β1 to β4). Remarkably, β1 and β3 and β2 and β4 respectively share a sequence similarity suggesting that the HET-s PFD was generated by an ancient duplication event. The proposed model corresponds to a pseudo-dimer structure with a double β-strand-turn-β-strand motif connected by a large loop (Fig. 3). Proline substitutions in any of the 4 β-strands abolish or strongly reduce infectivity and aggregation indicating that the ability to form the β-fold is critical for infectivity. This correlation between structure and infectivity strongly suggests that the proposed structure corresponds to the infectious fold of the protein. In contrast to many other amyloids, the solid state NMR spectra of HET-s(218-289) amyloids are characterized by extremely narrow peaks which reflects a very high level of structural organisation in the β-strand regions.[23,24]

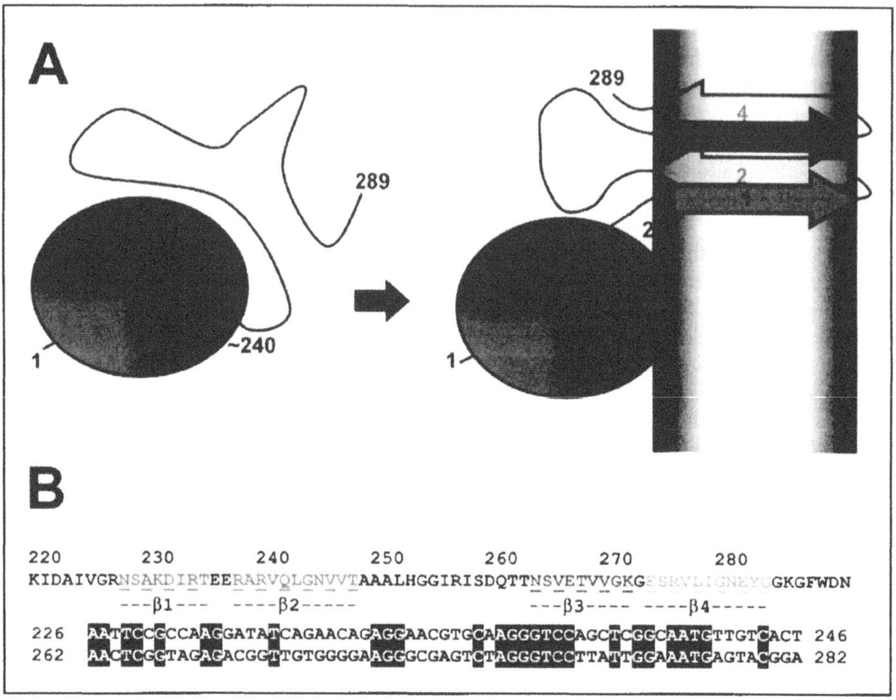

Figure 3. Domain organisation and structural model of HET-s. In A, a diagram of the HET-s domain organisation is given. In its soluble form, the protein displays a N-terminal α-helical globular domain spanning approximately residue 1 to 240 followed by a flexible C-terminal tail. In the amyloid form, the C-terminal region adopts a β-sheet rich conformation composed of 4 β-strands. The grey shading represents the fibril core. In B, the sequence of the HET-s PFD is given. The 4 β-strands are coloured. Below an amino acid and nucleic acid alignment of the β1-β2 and β3-β4 regions is given. Numbering correspond to the amino acid position in the HET-s sequence.

An Infectious Protein Encoded by an Invasive Allele

In 1965, Jean Bernet yet another former pupil of Georges Rizet described a remarkable property of [Het-s] in the sexual cycle.[25] Georges Rizet had shown that in a sexual cross of a prion infected [Het-s] strain with *het-S,* the prion is lost in the meiotic progeny. For this reason, he stated that the HET-S factor has the ability to inactivate [Het-s] (again this explains the original designation of [Het-s*]: "s^{S}" that should be understood as "*s* modified by *S*"). Now, when a [Het-s] x [Het-S] cross is performed at low temperature (18° instead of the usual 25°) and when [Het-s] is used as maternal parent something remarkable happens.[25,26] A proportion of the asci (about 20 %) are two spored instead of four spored. In these two spored asci, the remaining two spores are aborted. Bernet found that the two surviving spores are always of the *het-s* genotype indicating that the killed spores have the *het-S* genotype. In addition, in two spored asci the *het-s* spores contain [Het-s] but in four spored asci, the *het-s* spores produce [Het-s*] mycelia. No two spored asci are observed in a [Het-s*] X [Het-S] cross.

This *het-S*-spore killing is though to result from the toxic [Het-s]/HET-S interaction. In other words, when the HET-S protein is expressed in the maturating *het-S* spores, the interaction with the [Het-s] prion originating from the maternal cytoplasm leads to abortion of the *het-S* spores. Spore killing is asymmetric, in the reciprocal cross with the [Het-s] strain as the paternal parent there is no occurrence of spore killing. The male gametes (termed microconidia) contribute very little if any cytoplasm to the zygote, as a result the prion does not enter the sexual cycle when carried by the paternal parent. The percentage of killing varies during maturation of the fruiting body. Up to 60% of the asci in young perithecia show spore killing while in the older perithecia only 5% of the asci show killing. During the asci formation, the transition from coenocytic to cellular state results in a stochastic distribution of [Het-s] particles. Lowering of killing percentages with perithecial age, can be explained by hypothesizing that the young asci inherit most of the prion aggregates from the maternal cytoplasm. If the *het-s* expression stops during the sexual cycle, the late asci would inherit fewer prion seeds and thus killing percentages would diminish.[27] An alternative hypothesis, is that the progressive build up of the HET-S protein in the ascogonium during perithecial maturation leads to curing of the [Het-s] prion in late asci. Indeed, it is important to note that in this cross, the majority of the *het-s* ascospores are cured from the prion (even though the maternal parent was [Het-s]), this loss of [Het-s] indicates that *het-S* exerts a prion curing effect in a large fraction of the asci.

Spore killing efficiency can be highly increased (>80% killing) and rendered independent of the temperature by over-expressing the *het-s* gene.[26] The HET-s PFD is strictly required for the spore-killing activity.[27] The spore-killing activity empowers the *het-s* allele with the ability to be over-represented over *het-S* in the progeny of a *het-s x het-S* cross; *het-s* has thus the ability to cheat with Mendelian segregation and behaves as a meiotic drive element. Meiotic drive elements are ultra-selfish chromosomal genes that are genetically invasive and can reach high levels in populations without conferring any fitness benefit to the organism that harbors them.[28,29] Meiotic drive activity of the *het-s* allele is a direct consequence of the prion behaviour of HET-s. In other words, the *het-s* allele is invasive because it encodes an infectious protein. Why then are there still *het-S* alleles around ? Why hasn't *het-s* reach fixation ? One can envision a number of reasons for that. First, maybe *het-s* is on its way to fixation and Rizet arrived just in time to discover [Het-s]. Indeed without *het-S* alleles, [Het-s] would have never been discovered. Then, *het-S* might have a cellular function and thus a direct fitness advantage over *het-s* that would compensate for the meiotic drive effect. Third, if the biological function of the *het-s/het-S* system is somatic nonself recognition then these alleles are expected to be under balancing selection (i.e., equilibrated frequencies of the two alleles would be selected for thus preventing fixation of either one of them). Finally and maybe most importantly, is the battle occurring in the [Het-s] X *het-S* outcross, if the *het-S* side clearly loses on the genetic ground, *het-S* still has an epigenetic revenge as it leads to inactivation of the prion in most asci. This prion-curing effect of *het-S* might potentially greatly decrease the efficiency of *het-s* meiotic drive in natural populations and might represent the only natural situation of prion-curing in the wild.

Concluding Remarks, the Road Ahead...

More than half a century after its discovery, [Het-s] has unveiled part of its mystery. Thanks to the careful pioneering work of Georges Rizet and Janine Beisson and the molecular characterisation of the gene carried out by Joel Bégueret, the prion concept could be applied to this system. It is now clear that the [Het-s] genetic element is a protein. Recent work on [Het-s] and convergent evidence from the yeast prion field strongly suggest that [Het-s] corresponds to an amyloid or amyloid-like β-sheet-rich aggregated form of the HET-s protein. Nevertheless, as for all other prions, the exact nature of the entity propagating in vivo is really not known. Similarly, the mechanism of spontaneous prion appearance is not understood. In addition to these central questions on prion propagation which are common to all systems, [Het-s] also holds a number of more private secrets. Questions regarding the mechanism of incompatibility (why is the HET-S/HET-s interaction toxic?) as well the evolutionary history of *het-s* remain totally unanswered. The "petit *s*" element now stands in this *chiaroscuro*, with part of its mystery in the light and many more puzzles still hidden in the dark.

References

1. Rizet G. Les phénomènes de barrage chez Podospora anserina. I. Analyse de barrage entre les souches s et S. Rev Cytol Biol Veg 1952; 13:51-92.
2. Glass NL, Kaneko I. Fatal attraction: Nonself recognition and heterokaryon incompatibility in filamentous fungi. Eukaryot Cell 2003; 2(1):1-8.
3. Buss LW. Somatic cell parasitism and the evolution of somatic tissue compatibility. Proc Natl Acad Sci USA 1982; 79(17):5337-5341.
4. Saupe SJ. Molecular genetics of heterokaryon incompatibility in filamentous ascomycetes. Microbiol Mol Biol Rev 2000; 64(3):489-502.
5. Beisson-Schecroun J. Incompatibilité cellulaire et interactions nucléocytoplamsiques dans les phénomènes de barrage chez le Podospora anserina. Ann Genet 1962; 4:3-50.
6. Belcour L. Loss of a cytoplasmic determinant through formation of protoplasts in Podospora. Neurospora Newslett 1976; 23:26-27.
7. Turcq B, Denayrolles M, Béguerct J. Isolation of two allelic incompatibility genes s and S of the fungus Podospora anserina. Curr Genet 1990; 17:297-303.
8. Turcq B, Deleu C, Denayrolles M et al. Two allelic genes responsible for vegetative incompatibility in the fungus Podospora anserina are not essential for cell viability. Mol Gen Genet 1991; 228(1-2):265-269.
9. Deleu C, Clave C, Begueret J. A single amino acid difference is sufficient to elicit vegetative incompatibility in the fungus Podospora anserina. Genetics 1993; 135(1):45-52.
10. Deleu C, Turcq B, Begueret J. Repa, a repetitive and dispersed DNA sequence of the filamentous fungus Podospora anserina. Nucleic Acids Res 1990; 18(16):4901-4903.
11. Coustou V, Deleu C, Saupe S et al. The protein product of the het-s heterokaryon incompatibility gene of the fungus Podospora anserina behaves as a prion analog. Proc Natl Acad Sci USA 1997; 94(18):9773-9778.
12. Wickner RB. [URE3] as an altered URE2 protein: Evidence for a prion analog in Saccharomyces cerevisiae [see comments]. Science 1994; 264(5158):566-569.
13. Satpute-Krishnan P, Serio TR. Prion protein remodelling confers an immediate phenotypic switch. Nature 2005; 437(7056):262-265.
14. Derkatch IL, Bradley ME, Hong JY et al. Prions affect the appearance of other prions: The story of [PIN(+)]. Cell 2001; 106(2):171-182.
15. Coustou-Linares V, Maddelein ML, Begueret J et al. In vivo aggregation of the HET-s prion protein of the fungus Podospora anserina. Mol Microbiol 2001; 42(5):1325-1335.
16. Dos Reis S, Coulary-Salin B, Forge V et al. The HET-s prion protein of the filamentous fungus Podospora anserina aggregates in vitro into amyloid-like fibrils. J Biol Chem 2002; 277(8):5703-5706.
17. Maddelein ML, Dos Reis S, Duvezin-Caubet S et al. Amyloid aggregates of the HET-s prion protein are infectious. Proc Natl Acad Sci USA 2002; 99(11):7402-7407.
18. Wickner RB, Taylor KL, Edskes HK et al. Prions: Portable prion domains. Curr Biol 2000; 10(9):R335-337.
19. Balguerie A, Dos Reis S, Ritter C et al. Domain organization and structure-function relationship of the HET-s prion protein of Podospora anserina. EMBO J 2003; 22(9):2071-2081.
20. Nazabal A, Maddelain ML, Bonneu M et al. Probing the structure of the infectious amyloid form of the prion forming domain of HET-s using high-resolution hydrogen/deuterium exchange monitored by mass spectrometry. J Biol Chem 2005.
21. Nazabal A, Dos Reis S, Bonneu M et al. Conformational transition occurring upon amyloid aggregation of the HET-s prion protein of Podospora anserina analyzed by hydrogen/deuterium exchange and mass spectrometry. Biochemistry 2003; 42(29):8852-8861.
22. Balguerie A, Dos Reis S, Coulary-Salin B et al. The sequences appended to the amyloid core region of the HET-s prion protein determine higher-order aggregate organization in vivo. J Cell Sci 2004; 117(Pt 12):2599-2610.
23. Ritter C, Maddelein ML, Siemer AB et al. Correlation of structural elements and infectivity of the HET-s prion. Nature 2005; 435(7043):844-848.
24. Siemer AB, Ritter C, Ernst M et al. High-resolution solid-state NMR spectroscopy of the prion protein HET-s in its amyloid conformation. Angew Chem Int Ed Engl 2005; 44(16):2441-2444.
25. Bernet J. Mode d'action des gènes de barrage et relation entre l'incompatibilité cellulaire et l'incompatibilité sexuelle chez le Podospora anserina. Ann Sci Natl Bot 1965; 6:611-768.
26. Dalstra HJ, Swart K, Debets AJ et al. Sexual transmission of the [Het-S] prion leads to meiotic drive in Podospora anserina. Proc Natl Acad Sci USA 2003; 100(11):6616-6621.
27. Dalstra HJ, van der Zee R, Swart K et al. Non-Mendelian inheritance of the HET-s prion or HET-s prion domains determines the het-S spore killing system in Podospora anserina. Fungal Genet Biol 2005; 42(10):836-847
28. Lyttle TW. Cheaters sometimes prosper: Distortion of mendelian segregation by meiotic drive. Trends Genet 1993; 9(6):205-210.
29. Pennisi E. Meiotic drive. Bickering genes shape evolution. Science 2003; 301(5641):1837-1839.

CHAPTER 4

Prion-Prion Interactions

Irina L. Derkatch* and Susan W. Liebman

Abstract

The term prion has been used to describe self-replicating protein conformations that can convert other protein molecules of the same primary structure into its prion conformation. Several different proteins have now been found to exist as prions in *Saccharomyces cerevisiae*. Surprisingly, these heterologous prion proteins have a strong influence on each others' appearance and propagation, which may result from structural similarity between the prions. Both positive and negative effects of a prion on the de novo appearance of a heterologous prion have been observed in genetic studies. Other examples of reported interactions include mutual or unilateral inhibition and destabilization when two prions are present together in a single cell. In vitro work showing that one purified prion stimulates the conversion of a purified heterologous protein into a prion form, suggests that facilitation of de novo prion formation by heterologous prions in vivo is a result of a direct interaction between the prion proteins (a cross-seeding mechanism) and does not require other cellular components. However, other cellular structures, e.g., the cytoskeleton, may provide a scaffold for these interactions in vivo and chaperones can further facilitate or inhibit this process. Some negative prion-prion interactions may also occur via a direct interaction between the prion proteins. Another explanation is a competition between the prions for cellular factors involved in prion propagation or differential effects of chaperones stimulated by one prion on the heterologous prions.

Introduction

Originally, the term prion was coined to stress the "protein only" nature of an infectious agent causing transmissible spongiform encephalopathies in mammals.[1] All variations of this disease in humans and animals were linked to abnormal self-propagating conformations of just one cellular protein, PrP. The intriguing but controversial idea that a protein conformation could be infectious gained considerable support in 1994 when Reed Wickner showed that the prion model could explain the inheritance and behavior of the yeast [URE3] cytoplasmic factor and postulated that another yeast cytoplasmic factor, [*PSI*$^+$], was also a prion.[2] Soon, the discovery that the prion form of [Het-s] has a defined cellular activity in the fungus *Podospora anserina*, indicated that prions could have functional roles,[3,4] whereas the determination that the Hsp104 chaperone is required for the propagation of all known yeast prions showed that cellular machinery has a role in prion replication.[5-8] Now the protein only nature of these fungal cytoplasmic determinants has been definitively proven by demonstrating the infectivity of prion-like particles made in vitro from respective recombinant proteins.[9-12]

The list of prions and the number of prion-carrying species has expanded, but *Saccharomyces cerevisiae* is the only species in which several prions have been identified: [*PSI*$^+$], [URE3] and [*PIN*$^+$] (also known as [*RNQ*$^+$]). This makes yeast uniquely suited to study interactions

*Corresponding Author: Irina L. Derkatch—Department of Microbiology, New York University School of Medicine, NYU Medical Center, 530 First Avenue, New York, New York 10016, U.S.A. Email: irina.derkatch@med.nyu.edu

between different prions. Here we review the experimental evidence for these interactions, consider models to explain their molecular basis and briefly discuss their functional and evolutionary significance.

Structural Similarity between Prions Determines the Principle Types of Prion-Prion Interactions

The different proteins known to be capable of forming prions are not homologous. Although the prion forming domains of several prion proteins (Sup35, Ure2, Rnq1) are rich in glutamines (Q) and/or asparagines (N), other prion proteins (Het-s, PrP) are not Q- or N-rich. When not in their prion conformations, these proteins perform very different jobs in the cell. For example, the [URE3]-forming protein, Ure2, regulates the uptake of nitrogen (see ref. 13 for review). Sup35, which can convert into the [*PSI*⁺] prion, is normally a GTPase subunit of the translational termination factor responsible for the release of nascent protein chains from the ribosome (see ref. 14 for review). Het-S/[Het-s] is a determinant for heterokaryon incompatibility (see ref. 15 for review) and selective spore killing in *Podospora anserina*.[16] These functions are clearly distinct from the roles assigned to cellular PrP in the central nervous system (see refs. 17,18 and refs. therein).

However, despite the dissimilarity of their building blocks, the above mentioned prions are strikingly alike (see ref. 19 for review). Their prion aggregates are β-sheet-rich, i.e., prion formation is associated with an increase of β-sheet content relative to non-prion states of the same protein. The joining of new molecules occurs through the formation of inter-molecular interactions between β-strands. Consequently, unlike disordered amorphous aggregates that are generally only held together by hydrophobic interactions, prion aggregates are highly ordered. This explains the poor solubility of prion aggregates in detergents and their resistance to digestion with proteases. Whereas the detailed in vivo architecture of prion aggregates is unknown, the overall arrangement of proteins in prion aggregates strongly resembles that found in amyloid fibers. Indeed, prions bind the Thioflavin T and Congo Red dyes the same way other amyloids do.[20,21] It has also been established that larger aggregates, which can be detected by gradient centrifugation and even light microscopy, are composed of smaller SDS-resistant subparticles.[22,23] These subparticles could correspond to units of prion propagation and are sometimes referred to as seeds.

In vitro, prion-forming proteins assemble into typical amyloid fibers, a process which can be self-seeded by preformed fibers of the same protein and by extracts of cells bearing the respective prion.[8,24-30] Also, both PrP and Sup35 have been shown to form oligomers that have a common structure and that precede the formation of amyloid fibers.[31,32] These oligomers are recognized by an antibody that reacts with similar oligomers formed by a wide variety of amyloid-forming proteins including Aβ and polyglutamine.[31]

Therefore, prions are amyloid aggregates that are heritable and transmissible. Specifically for yeast, this implies that prion-forming proteins have a propensity to form amyloid aggregates that can be partitioned and transmitted during cell divisions. Below we consider how the common properties of these prions may determine the framework and nature of prion-prion interactions.

Direct Interaction

Firstly, since prions are amyloids, it is possible that molecules of one prionogenic protein could join a heterologous prion aggregate. Indeed, the basic structure of all amyloid aggregates is the same (for review see ref. 33): a ladder of β-strands oriented perpendicular to the fibril axis that may be organized in β-sheets, β-helices, β-nanotubes, etc. The bonds in the β-strands are formed between the peptide backbones, not between the side chains. Thus, prionogenic proteins, prone to form β-strands with amyloids of the same primary sequence, might occasionally attempt to form a β-strand with heterologous amyloids, as long as the side chains do not interfere with the backbone interactions. This would likely block the growth of the fiber to which such a heterologous "cap" was attached, but could simultaneously lead to the de novo formation of a new prion by a cross-seeding mechanism.

Competition for Cellular Factors

The critical steps in prion biogenesis and propagation are the same for different amyloidogenic prions. Initially several functional cellular proteins form a prion nucleus or seed. The seed then grows by the addition of new molecules. Finally, the large prion aggregate is broken to create new seeds. At each of these steps the sets of cellular factors that inhibit and promote the formation and propagation of various prions are likely to overlap considerably (see refs. 34,35 for reviews) and competition for such factors between heterologous prions is likely. The effects of such competitions might depend upon the stage at which a particular factor affects prion formation and the particular role the factor plays. For example, chaperones that promote the proper folding of nascent polypeptides by eliminating misfolded or partially folded prionogenic intermediates might interfere with the biogenesis of prions. The titration of such factors could promote prion formation. On the other hand, chaperones that break amyloid aggregates into smaller seeds are expected to promote prion propagation. The titration of the chaperones could decrease prion stability. Also, because different prions respond differently to the depletion of various cellular factors,[36] their effects on each other may not be reciprocal.

Cellular Response to the Presence of Prions

Finally, the cell may mount a response to the presence of amyloid or amyloid precursors of prions.[36,37] This response is expected to be aimed at prion disaggregation and elimination but may actually facilitate prion transmission, e.g., by breaking prion aggregates into smaller seeds. Such a generic response would affect not only the prion protein that induced it, but other, heterologous prions and prionogenic proteins.

In the following sections we describe known examples, where different prions affect de novo formation, maintenance and phenotypic expression of each other.

Prions Facilitate the de Novo Appearance of Heterologous Prions

[PIN$^+$]: *the First Example of Prion-Prion Interactions*

The idea that the de novo formation of prions is facilitated by preexisting heterologous prion aggregates is linked to the discovery of [*PIN$^+$*].[6,38,39]

[*PIN$^+$*]: a Prion Interacting with [*PSI$^+$*]

As prions are alternative self-propagating conformations of cellular proteins, they should occasionally appear in cell populations, and the spontaneous loss or curing of a prion should not preclude the possibility of its reappearance.[2,40] Indeed, the rare spontaneous appearance of [URE3] was demonstrated both in [ure^0] strains and in derivatives of [URE3] strains treated with the most conventional prion-curing agent, guanidine hydrochloride (GuHCl).[2,41] Spontaneous de novo appearance of [*PSI$^+$*] was also reported[42] but evidence for its reappearance after curing with GuHCl remained controversial.[42,43]

The issue of the reappearance of [*PSI$^+$*] after its curing with GuHCl was revisited in the experimental system where the de novo formation of [*PSI$^+$*] could be induced 100-fold or more by a 5 to 10 fold overproduction of the Sup35 protein.[44,45] Such an increase in the rate of [*PSI$^+$*] prion formation upon overproduction of the prion-forming protein is predicted by the prion model because the accumulation of Sup35 molecules means there is a greater chance that a group of them will misfold and form a prion nucleus.[2,40] Also, the dramatic increase in [*PSI$^+$*] formation can be attributed to the overloading of the protein folding machinery and the imbalance of overexpressed Sup35 relative to its normal binding partners.[46,47] To our surprise we eventually established that [*PSI$^+$*] can be induced (or reappear spontaneously) only in a fraction of GuHCl-cured [*psi$^-$*] derivatives.[6] The proportion of nonrevertible clones increased with increasing GuHCl treatment, but the efficiency of [*PSI$^+$*] induction in the clones that remained inducible was unchanged. Thus a GuHCl-curable factor was required for [*PSI$^+$*] formation. The factor was named [*PIN$^+$*] for [*PSI$^+$*] induction.[6]

Genetic analysis of the Pin+ phenotype established that [*PIN*+] is inherited in a non-Mendelian fashion. Diploids resulting from crosses of [*PIN*+] and [*pin*-] derivatives were always [*PIN*+], and in meiosis the segregation was 4 [*PIN*+] : 0 [*pin*-].[6] Also, [*PIN*+] was efficiently transmitted by cytoduction,[38] an abortive mating where cytoplasms of two cells blend but their nuclei do not fuse resulting in heterokaryons, which bud off haploids with single unmodified parental nuclei and mixed cytoplasm.[48] Such inheritance is possible for (i) genetic determinants located on nonnuclear DNA or RNA molecules (plasmids, viruses or mitochondrial DNA), or (ii) for prions.[2]

Elimination of [*PIN*+] on GuHCl-containing media was compatible with both possibilities since GuHCl both cures prions and induces large deletions in mitochondrial DNA.[49,50] However, no link between [*PIN*+] and mitochondrial DNA was detected.[39] Also, [*PIN*+] reappeared spontaneously following GuHCl curing,[38,39] whereas a [*pin*-] caused by a large deletion would probably not be revertible. On the other hand, disruption of the gene encoding the Hsp104 chaperone previously reported to cure [*PSI*+] and [URE3][5,7] also eliminated [*PIN*+].[6] Thus, [*PIN*+] was hypothesized to be a prion, and the effect of [*PIN*+] on the appearance of [*PSI*+] became the first example of a prion-prion interaction.

[*PIN*+] Interacts with [*PSI*+] at the Step of [*PSI*+] Formation

Although essential for the appearance of [*PSI*+], [*PIN*+] is dispensable for [*PSI*+] maintenance: [*PSI*+][*pin*-] derivatives (along with [*psi*-][*PIN*+] and [*psi*-][*pin*-] derivatives) were obtained by growing [*PSI*+][*PIN*+] cultures on GuHCl media.[39] Furthermore, the loss of [*PIN*+] generally had no effect on [*PSI*+] properties such as the size of [*PSI*+] prion subparticles[23] or the degree of Sup35 aggregation.[39] Consequently, the [*PSI*+] phenotype of nonsense suppression was unaffected.[39] This phenotype is caused by depletion of soluble Sup35 available for the termination of translation and is routinely analyzed using the *ade1-14* (UGA) or *ade2-1* (UAA) reporters with a premature stop codon.[43,51] [*PSI*+] toxicity upon Sup35 overproduction,[52,53] which may result from impairment of the cytoskeleton,[54] was also unaffected by [*PIN*+].[6] Only the inhibition of growth in [*psi*-] upon high-level Sup35 overexpression required the presence of [*PIN*+], but this phenotype is attributed to the induction of [*PSI*+].[6] Taken together these data indicate that [*PIN*+] specifically facilitates the de novo formation of [*PSI*+].

Rnq1 Is the [*PIN*+] Prion Protein

The prion model postulates three approaches for identifying a prion protein: (i) disruption of the gene encoding the prion protein should eliminate the respective prion; (ii) overexpression of the prion protein gene should promote de novo prion formation and (iii) prion formation should inhibit the normal function of the prion-forming protein.[40]

Rnq1 was identified as the [*PIN*+] prion protein using a candidate approach. Since disruption of the gene encoding the [*PIN*+] protein should cure cells of [*PIN*+], genes for all known and a few candidate prions were disrupted in a [*PIN*+] strain. Strains with *ure2*Δ, *new1*Δ, *pin2*Δ and *sup35-NM*Δ (the latter lacking the prion domain of the otherwise essential *SUP35* gene) remained [*PIN*+].[6,38] However disruption of the *RNQ1* gene caused the [*PIN*+] strain to become Pin- and all attempts to cytoduce [*PIN*+] from the *rnq1*Δ background into a wild type *RNQ1* strain were unsuccessful.[38] Newly appearing spontaneous [*PIN*+]s also could be cytoduced into a wild type *RNQ1* strain but not into *rnq1*Δ.[38] Furthermore, the presence of [*PIN*+] correlated with an aggregated state of the Rnq1 protein. In [*PIN*+] strains Rnq1 was insoluble; it became soluble in [*pin*-] clones obtained on GuHCl and converted back to an aggregated state following spontaneous [*PIN*+] reappearance.

RNQ1 encoded a prion previously identified in laboratory yeast strains[8] and recently found in wild type yeast.[55,56] It was discovered by Sondheimer and Lindquist,[8] who noted a striking similarity between the C-terminal domain of Rnq1 and the prion domain of Sup35. The similarity was manifested by an extremely high frequency of Q/N residues, which gave the gene its name (rich in N and Q). This C-terminal region was sufficient to maintain the prion (aggregated) state of Rnq1 and was thus confirmed to be the prion domain of [*RNQ*+].[8] The same prion was also sufficient for the propagation of [*PIN*+].[38] Thus, [*RNQ*+]=[*PIN*+].

A Broader Ramification of the [PIN$^+$] Phenomenon: Various Heterologous Prions Affect Each Other's Appearance

Other Prions Also Have Pin$^+$ Activity

Another approach to identifying the [*PIN$^+$*] prion protein led to a very different conclusion. Since, according to the prion model, overexpression of the [*PIN$^+$*] protein gene would promote the de novo formation of [*PIN$^+$*], a multicopy library of yeast genomic DNA was introduced into a [*pin$^-$*] derivative to screen for yeast genes that, upon overexpression, caused cells to become Pin$^+$ (i.e., inducible to [*PSI$^+$*]). Unexpectedly, the screen did not uncover *RNQ1* but instead yielded 11 genes encoding functionally unrelated proteins with domains unusually rich in Q/N.[38] Among them were *URE2* and *NEW1*, the genes for a well-known prion [URE3] and an artificial prion [*NU$^+$*], both of which had already been eliminated as the [*PIN$^+$*] protein encoding genes in the candidate approach (see above). The ability of *NEW1* overexpression to make cells Pin$^+$ was simultaneously reported in a separate study.[57] Furthermore, the Pin$^+$ phenotype of facilitating the de novo formation of [*PSI$^+$*] was not only caused by overexpression of Ure2 and New1, but also by the presence of the [URE3] or [*NU$^+$*] prions.[38,57] Thus, [URE3] and [*NU$^+$*] give cells a Pin$^+$ phenotype. Indeed, to date there is no known yeast prion that does not make cells Pin$^+$.

Positive Effects of [URE3], [*PSI$^+$*] and [*PIN$^+$*] on Each Other's Appearance

[*PIN$^+$*] itself is strongly dependant upon other prions for its formation. Indeed, although [*PIN$^+$*] can appear spontaneously in cultures devoid of any known prions, this is an extremely rare event.[39] Also, in the absence of other prions the frequency of fluorescent aggregates caused by overexpression of an Rnq1-GFP fusion construct (indicative of prion formation) was only ~0.1%, even after prolonged overexpression of the construct. However, the presence of [URE3] or [*PSI$^+$*] led to up to a 100-fold increase in the formation of bright fluorescent Rnq1-GFP foci.[38] Thus, poor induction of [*PIN$^+$*] by *RNQ1* overexpression in prion-less strains explains why *RNQ1* was not identified in the library screen for the [*PIN$^+$*] maintenance gene described above.

[URE3] is similar to [*PIN$^+$*] in its ability to form in GuHCl-cured cultures devoid of any known prions. And, as in the case of [*PIN$^+$*], the frequency of [URE3] appearance in such cultures is extremely low: ~10^{-5} spontaneously and ~10^{-3} following Ure2 overexpression. [*PIN$^+$*] increases the induction of [URE3] 5 to 500-fold.[58]

Negative Effects of [URE3] and [*PSI$^+$*] on Each Other's Appearance

While it is clear from the data discussed above that the de novo appearance of prions other than [*PSI$^+$*] can be promoted by the presence of heterologous prions, this is not always the case. Two groups found that [*PSI$^+$*] did not promote [URE3] formation but rather inhibited it.[36,58] Furthermore, in one study [URE3] had a negative effect on the induction of [*PSI$^+$*],[36] rather than the positive one described above.[38] Since Bradley et al[58] detected a positive effect of [*PIN$^+$*] and negative effect of [*PSI$^+$*] on appearance of [URE3] in the same genetic background, their result cannot be explained by a genomic mutation interfering with all positive prion-prion interactions. It is possible that for some pairs of prions the effect of heterologous facilitation is simply not reciprocal and that [*PSI$^+$*] never promotes the appearance of [URE3]. On the other hand, it is possible that a particular variant of [*PSI$^+$*] chosen for these experiments was incapable of promoting the de novo formation of [URE3]. The latter explanation is also compatible with detection of both negative and positive effects of [URE3] on [*PSI$^+$*].

Conformational Variants of Prions Differ in Their Propensity to Induce Heterologous Prions

Prions resulting from independent conversions of the same soluble proteins into amyloid aggregates are not all identical. For most prionogenic proteins there is an array of prion states called prion strains, or variants.[45,58-60] Prion variants are due to differences in the conformation of prion aggregates (see ref. 19 for review). Apparently, different lengths and arrangements of β-strands and/or different mutual orientations of β-sheets in the amyloid structure,[61-64] affect the rate of aggregation and the resistance of amyloid aggregates to fragmentation and, consequently, result in different prion seed sizes.[19,22,23]

Strong [*PSI*⁺] variants have smaller prion subparticles than weak [*PSI*⁺] variants but cause more aggregation of Sup35 and therefore cause more efficient nonsense suppression.[22,45,65] This suggests that there is a larger number of strong [*PSI*⁺] seeds relative to weak [*PSI*⁺] seeds per cell. When the effects of strong and weak [*PSI*⁺]s on the prionization of Rnq1 were compared, a strong [*PSI*⁺] variant caused the appearance of approximately two-fold more Rnq1-GFP foci than a weak [*PSI*⁺] variant following either short or prolonged overexpression of Rnq1-GFP.[38] However, because only two [*PSI*⁺] variants were compared, the correlation between the strength of [*PSI*⁺] and its ability to induce [*PIN*⁺] cannot be established.

Five [*PIN*⁺] variants that differed in the degree of Rnq1 aggregation also had different propensities to promote [*PSI*⁺] induction[58] (for review see ref. 66). There was no correlation between the frequency of [*PSI*⁺] induction and the degree of Rnq1 aggregation or the size of [*PIN*⁺] subparticles. The best [*PSI*⁺] induction was detected in the derivative with the most soluble Rnq1, whereas the derivative with the least soluble Rnq1 was ranked second in the [*PSI*⁺] induction test.[58] Furthermore, when these [*PIN*⁺] variants were screened for their ability to facilitate [URE3] formation, there was no reproducible difference between them, whereas yet another [*PIN*⁺] variant from the Wickner lab collection promoted [URE3] appearance much more efficiently.[58] Thus, prion variants facilitate the induction of heterologous prions to different extents, and this ability is apparently determined by their conformational differences.

Interactions with Non-Prion Amyloids

Interactions between polyQ Amyloids and Prions

Since the prion domains of [*PIN*⁺], [*PSI*⁺], [URE3] and [*NU*⁺] are QN-rich and all the proteins identified in the genetic screen described above also contain QN-rich domains, it seemed likely that their Pin⁺ activity resulted from the presence of QN-rich sequences. Another type of Q-rich sequences, uninterrupted polyQ stretches, is found in many proteins. When expanded beyond an acceptable limit, polyQ stretches are prone to form β-sheet-rich amyloid aggregates, which are associated with several human diseases including Huntington's disease and MJD. Several groups utilized constructs expressing polyQ-expanded fragments of huntingtin (Ht) and MJD for the analysis of interaction between yeast prions and aggregation-prone polyQ.

As expected, overexpression of constructs carrying the first exon of Ht with expanded polyQ, rendered a fraction of [*pin*⁻] cells phenotypically Pin⁺, and the ability of Ht constructs to promote [*PSI*⁺] induction correlated with their ability to aggregate.[67] In a reverse experiment [*PIN*⁺] and [*NU*⁺], but not [*PSI*⁺], promoted aggregation of polyQ-expanded MJD.[57] Also, [*PIN*⁺] was shown to promote aggregation and toxicity of the polyQ-expanded Ht fragments,[68] and recent evidence suggests that [*PSI*⁺] has the same effect on the Ht toxicity as [*PIN*⁺].[69] Thus, non-prion polyQ amyloids can engage in interactions with prion proteins at the step of de novo amyloid formation. The fact that MJD aggregation was not stimulated by [*PSI*⁺],[57] confirms that such interactions are not obligatory and may be nonreciprocal.

Interactions between Non-polyQ Amyloids and Prions

The question about interactions between Q/N- rich and non-Q/N-rich prions remains open. Perutz explained the high propensity of Q-, N- and Q/N-rich sequences to form amyloid by the involvement of Q and N residues in the stabilization of β-strands.[70,71] One possibility is that there are interactions that are specific to Q/N-rich amyloids. Indeed, only Q/N rich proteins were identified as facilitators of the appearance of [*PSI*⁺] in the genetic screen (see above). Also, the formation of [*PSI*⁺] was not facilitated by non-Q- or N- rich aggregates of the normal and disease variants of the amyloidogenic proteins transthyretin, α-synuclein and synphilin-1.[67] However, it should be recalled that even in the case of the Q/N-rich amyloids, not every pair of amyloids can seed each other. Thus it is possible that most amyloidogenic and prionogenic proteins have Q/N-rich domains, and interactions with less common non-polyQ amyloids will be eventually established.

Interactions at the Step of the de Novo Prion Formation Can Be Reproduced in Vitro

The assembly of prion proteins into amyloid fibers in vitro is fundamentally similar to prion formation and growth in vivo. Following a lag period, a pure recombinant protein can form fibers in vitro. The lag period can be reduced or eliminated by the addition of preformed fibers made of this protein or by adding extracts from cells bearing a prion form of this protein.[8,24-30] Furthermore, amyloid prepared in vitro from pure recombinant proteins can efficiently transform live cells to the prion state.[9-12,72,73] Most phenomena associated with prion biogenesis and propagation, including the effects of chaperones[32,74-76] and the existence of strain phenomenon and species barriers[62,77-82] have been modeled in vitro using purified recombinant prionogenic proteins and cellular factors.

In the two-step process of amyloid formation in vitro, the lag phase before the ThT or Congo Red-binding fibers are detected is equivalent to the step of the de novo prion formation in vivo. Subsequent rapid growth of fibers models the joining of existing prion aggregates by newly synthesized protein molecules. Accordingly, events equivalent to the prion-prion interactions at the step of the de novo formation should affect the length of the lag. Indeed, addition of preformed Rnq1 fibers to soluble Sup35 shortened the lag before Sup35 amyloid could be detected.[67] The effect of Rnq1 fibers was very modest relative to the effect of the addition of homologous, Sup35 fibers, which is consistent with the ability of [*PIN*$^+$] to promote [*PSI*$^+$] appearance in only a fraction of the cells in the culture. As recombinant purified Rnq1 and Sup35 were used in these experiments, this result indicates a direct interaction between preexisting prion aggregates and a heterologous prionogenic protein. The non-polyQ amyloids that didn't facilitate [*PSI*$^+$] formation in vivo also didn't promote Sup35 formation in vitro, which further validates the use of the in vitro model for studies of prion-prion interactions. Interestingly, two non-poly Q amyloids, bovine pancreas insulin and human Ig light-chain amyloid, did stimulate Sup35 conversion in vitro.[67] Although these amyloids were not tested in vivo, this finding suggests that the presence of a QN-rich domain is not an absolute requirement for such interactions.

Prions Interfere with Propagation of Heterologous Prions

There are only a few examples of prion-prion interactions that affect the propagation of prions.

[URE3] and [PSI$^+$] Negatively Affect Each Other's Propagation

Schwimmer and Masison[36] showed that, at least in the [*PIN*$^+$] background they used, there are antagonistic interactions between [URE3] and [*PSI*$^+$]. A negative effect of [URE3] on [*PSI*$^+$] was detected in an *ade2-1* (UAA) *SUQ5* strain,[51] where [*PSI*$^+$] restores the red color caused by the *ade2-1* mutation to pink or white, depending upon the efficiency of nonsense suppression. [URE3] reduced the suppressor phenotype associated with [*PSI*$^+$]: [URE3][*PSI*$^+$] cultures were pink, whereas isogenic [ure^0][*PSI*$^+$] derivatives carrying the same [*PSI*$^+$] variant formed white colonies. The level of Sup35 aggregation in [URE3][*PSI*$^+$] cells was also lower than in [ure^0][*PSI*$^+$], which confirmed that [URE3] indeed inhibited [*PSI*$^+$] aggregation and did not interfere with the termination of translation in some other way. Furthermore, the inhibitory effect of [URE3] toward [*PSI*$^+$] increased when cells were grown on media selective for [URE3], although even under these conditions [*PSI*$^+$] remained mitotically stable. [*PSI*$^+$] exerted a reciprocal inhibitory effect on [URE3], which was detected as reduced de-repression of the *DAL5* promoter, due to the appearance of functional Ure2.[36] The impairment of [URE3] propagation by [*PSI*$^+$] was also apparent from the loss of [URE3] in 1% of the mitotically growing [*PSI*$^+$] derivatives.

*Incompatible Variants of [*PIN$^+$*] and [*PSI$^+$*]*

The next example of antagonistic prion-prion interactions came as a surprise. The [*PIN$^+$*] variant, which was originally detected in a laboratory strain[6] and established as a prion essential for the de novo formation of [*PSI$^+$*], had already been shown not to affect [*PSI$^+$*] propagation.[38] However, Bradley and Liebman[83] found that variants of [*PIN$^+$*] that appeared spontaneously following prolonged storage of GuHCl-cured [*pin$^-$*] derivatives[38] and were classified as low, medium and very high according to their ability to facilitate [*PSI$^+$*] formation,[58,66] destabilized weak variants of [*PSI$^+$*]. When weak [*PSI$^+$*] was introduced into these [*PIN$^+$*] derivatives by mating or cytoduction, approximately 50% of the cells in the resulting [*PIN$^+$*] colonies were [*psi$^-$*].[83] Strikingly, these "destabilizing" [*PIN$^+$*] variants even inhibited the propagation of the same weak [*PSI$^+$*] variants whose appearance they promoted at the step of induction.

As in the example described for [*PSI$^+$*] and [URE3] in the previous section, the antagonism of weak [*PSI$^+$*] and medium destabilizing [*PIN$^+$*] was reciprocal. This [*PIN$^+$*] frequently disappeared following the induction of weak [*PSI$^+$*] and if it was not lost right away, the [*PIN$^+$*][unstable weak *PSI$^+$*] derivative soon segregated into two derivatives, either [*PIN$^+$*][*psi$^-$*] or [*pin$^-$*][stable weak *PSI$^+$*]. On the contrary, low destabilizing [*PIN$^+$*] was not lost upon [*PSI$^+$*] induction and clearly won the competition with [*PSI$^+$*] thereafter. Thus, in the case of antagonistic prion-prion interactions, either one of the prions is eliminated from the population, or the population segregates into derivatives bearing different sets of prions.

How Might Heterologous Prionogenic Proteins Interact and What Might Be the Consequences?

Direct Interaction of Prion Aggregates with Heterologous Prionogenic Proteins

Cross-Seeding Model

The cross-seeding model (Fig. 1) explains the genetic phenomenon of [*PIN$^+$*] and is compatible with data on the facilitation of the de novo appearance of prions by preexisting heterologous prions or amyloids.[38,57] The model postulates that a preexisting aggregate can seed the formation of a new prion by providing a nidus for the assembly and/or conformational conversion during the early stages of prion biogenesis. For example, [*PIN$^+$*] aggregates are proposed to be used as the site of initial Sup35 assembly and conversion into [*PSI$^+$*].

A critical argument in favor of this model is the facilitation of the formation of Sup35 amyloid upon the addition of preformed Rnq1 fibers (see above).[67] This in vitro result, obtained using pure recombinant proteins, suggests a direct interaction between Rnq1 and Sup35, unmediated by other cellular factors.

The seeding model does not require a permanent interaction between the heterologous amyloids. Indeed, while the growth of amyloid aggregates by the addition of homologous proteins is a very efficient process,[84] the binding of a heterologous protein should occur much less frequently and should engage only a small fraction of prion propagons in the cell. Thus, the findings that in a [*PIN$^+$*][*psi$^-$*] strain, Sup35 remained monomeric and was not detected within Rnq1 prion subparticles, and that in a [*PIN$^+$*][*PSI$^+$*] strain the subparticles of Rnq1 and Sup35 were not intermixed[23] do not contradict the model.

The recent detection of mixed aggregates during [*PSI$^+$*] induction supports the model: when Sup35 was overexpressed in a [*PIN$^+$*][*psi$^-$*] derivative, newly forming detergent-insoluble Sup35 aggregates contained some Rnq1.[85] Also, using Rnq1 and Sup35 respectively labeled with yellow and cyan fluorescent proteins, we established that all newly forming [*PSI$^+$*] aggregates partially or completely co-localize with preexisting [*PIN$^+$*] aggregates.[67] This co-localization is hypothesized to be a consequence of the seeding event. Indeed, even though similar co-localization was also detected in [*PIN$^+$*][*PSI$^+$*] derivatives propagating two established prions, not all aggregates showed co-localization in these cultures.[67] This, latter, finding is compatible

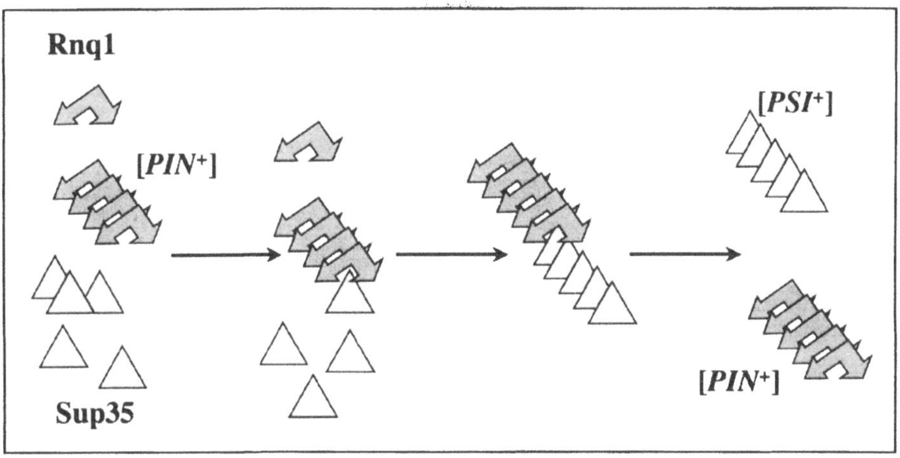

Figure 1. Cross-seeding model for [*PIN*+]. [*PIN*+] aggregates are proposed to be used as the site of initial Sup35 assembly and conversion into [*PSI*+] (Rnq1—blue arrows, Sup35—green triangles). A color version of this figure can be found at www.eurekah.com.

with the idea of occasional heterologous interactions between established prions, but can also be explained by the occasional lateral attachment of large amyloid structures or the co-compartmentalization of amyloids.

The fact that mutual induction is detected between proteins with similar Q/N-rich prion domains is compatible with the model. Although amyloid structure is determined by bonds formed by peptide backbones, side chains can either strengthen or weaken the structure.[33] One possibility is that Q and N residues form a hydrogen bond "polar zipper" that strengthens the β-spine.[71,86] Alternatively, the critical issue could be the interdigitation of side chains in the β-rich structure stabilizing it via hydrophobic interactions and allowing Van der Waals attractions favorable for amyloid formation to be maximized, while avoiding electrostatic repulsions.[71,87] When a short peptide derived from the Sup35 prion domain was used to model the structure of the cross-β-spine, Lipfert et al[88] found support for the polar zipper formation, whereas Nelson et al[63] predicted that Q and N side-chains were interdigitated to form a dry, tightly self-complementing steric zipper between two β-sheets. Whether prion amyloids are composed of polar or steric zippers, the positions of specific Q or N residues may be critical for efficient cross-seeding. Indeed, a single substitution in the Sup35 prion domain was reported to block its being seeded with wild type Sup35 amyloid without seriously interfering with the ability of the mutant to form amyloid in vitro in the absence of seed.[89]

Another line of support for the cross-seeding model comes from examples of interactions between non-prion amyloids. The mysterious amyloid enhancing factor (AEF), that reduces the time before amyloid protein A (AA) amyloidosis onset in chronic inflammation models, appears to be equivalent to fragments of amyloid fibrils. Strikingly, not only AA fibrils[90,91] but also fibrils of various human amyloids including transthyretin and islet amyloid polypeptide (IAPP),[92,93] Sup35,[94] or synthetic silk-derived fibrils,[95] had this effect. Overall, AA seeding appears to be rather nonspecific. On the contrary, the study of the specificity of seeding of Aβ fibers revealed that only IAPP was an efficient heterologous seeder for this peptide.[96] Strikingly, the authors noted the high similarity of the primary structures of Aβ and IAPP amyloidogenic peptides of otherwise nonhomologous proteins and attributed the cross-seeding to this feature.

Capping Model

Similar interactions between heterologous proteins could also lead to the curing of a preexisting prion by a heterologous prionogenic protein (Fig. 2). Indeed, binding of a heterologous

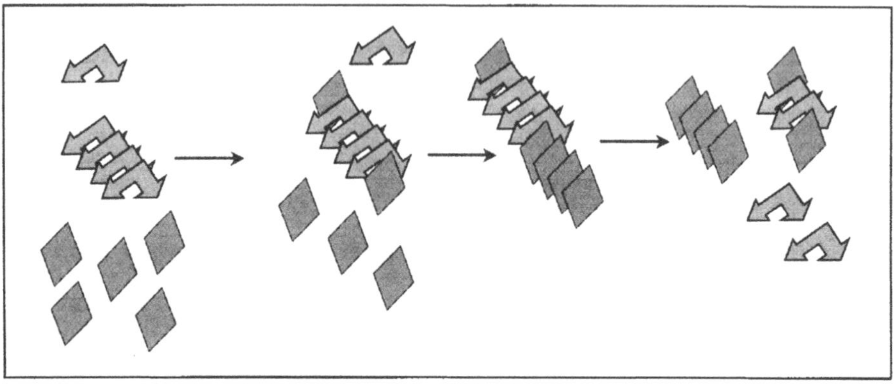

Figure 2. Capping model. The binding of a heterologous protein (magenta diamonds) to the growing tip of a prion aggregate (blue arrows) could block its rapid growth and lead to its destabilization and loss. At the same time such binding can lead to the formation of a prion by the heterologous protein by a seeding mechanism. A color version of this figure can be found at www.eurekah.com.

protein to the growing tip of a prion aggregate could block its rapid growth. Formation of such "caps" on a considerable fraction of prion particles, may lead to a notable reduction in the prion-associated phenotypes and to inefficient transmission to daughter cells.

A similar model was proposed to explain the Pnm ([*PSI*$^+$] no more) phenotype of Sup35 mutants towards wild type [*PSI*$^+$][89] and the curing of [URE3] upon overexpression of Ure2 fragments or GFP fusions from *S. cerevisiae* and other species.[97,98] The authors hypothesized that mutants or fragments would join the growth tip but would not themselves provide a growth point and would thus poison the amyloid crystal. Likewise, the elimination of some [*PIN*$^+$] variants upon overexpression of Sup35 and the induction of [*PSI*$^+$] could be explained by this model. However, destabilization of weak [*PSI*$^+$] by certain [*PIN*$^+$] variants in the absence of any overexpression is not easily explained by a capping model.[83]

Extending the Seeding Model: Role of Chaperones and Cytoskeleton

Although the in vitro evidence implies that the [*PIN*$^+$] effect is not strictly chaperone mediated,[67] it is still possible that in vivo chaperones bound to preexisting prion aggregates, rather than [*PIN*$^+$] per se, are mostly responsible for the effect on the de novo induction of [*PSI*$^+$] (Fig. 3).[99]

Also, newly appearing Sup35-GFP prion aggregates that appear in [*PIN*$^+$] cells, with the characteristic ring/ribbon shape that is easily distinguishable from aggregates of "established" prions,[38,100] have recently been reported to preferentially associate with actin patches.[54] It was thus hypothesized that actin patches provide a scaffold for the formation of large prion aggregates making these areas the sites of direct prion-prion interactions.[54]

Prion Interactions Mediated by Chaperones or Other Cellular Factors

Titration Model

This model (Fig. 4) was originally proposed to explain the [*PIN*$^+$] phenomenon.[38,57,99] The model postulates that cellular factors responsible for the disassembly of aggregates and/or refolding of misfolded proteins are constantly working to dissolve both existing and newly appearing prions. Thus, in the prion-free cells such chaperones efficiently prevent the appearance of new prions (Fig. 4 top). On the contrary, in the presence of another prion, disaggregating factors are titrated away to work on this prion. This allows newly forming prions to escape

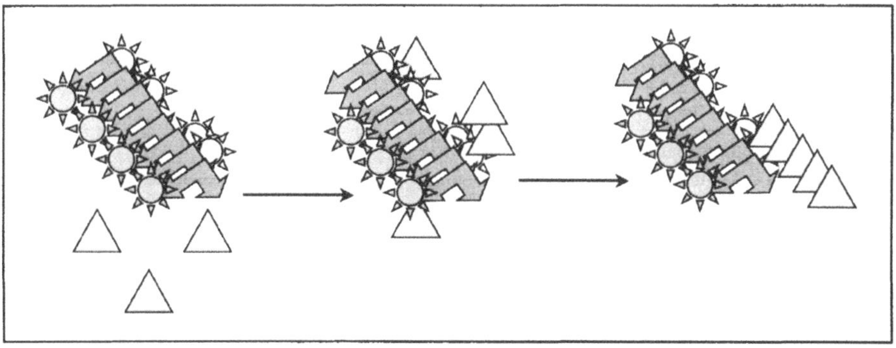

Figure 3. Chaperone-assisted seeding model for [*PIN*$^+$]. Chaperones (orange sun shapes) bound to preexisting [*PIN*$^+$] prion aggregates (blue arrows), could be responsible for the enhanced de novo aggregation of Sup35 (green triangles) and thus facilitate [*PSI*$^+$] formation in vivo. A color version of this figure can be found at www.eurekah.com.

the protein folding control machinery and achieve the stage of rapid propagation, which is resistant to chaperone curing.

While this model alone does not explain all of the experimental evidence for [*PIN*$^+$] and especially the in vitro reproduction of this phenomenon (see refs. 38,67,99 for discussion), the titration and the seeding models are not mutually exclusive. Indeed, the existence of a cellular factor with a weak inhibitory effect towards de novo aggregate formation was hypothesized by

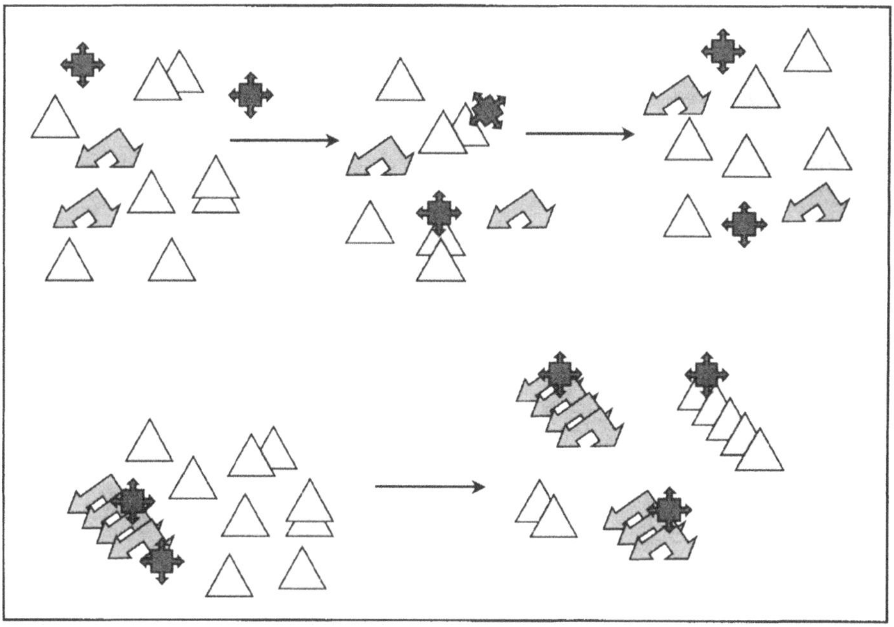

Figure 4. Titration model for [*PIN*$^+$]. Top: in a [*psi*$^-$] [*pin*$^-$] cell, cellular factors (purple shape) keep Sup35 (green triangles) and Rnq1 (blue arrows) from aggregating. Bottom: in a [*psi*$^-$] [*PIN*$^+$] cells, much of the factor is bound to the [*PIN*$^+$] aggregate, so less is available to keep Sup35 from aggregating. A color version of this figure can be found at www.eurekah.com.

Uptain et al[77] to explain their findings that extracts from [*psi*⁻][*pin*⁻] strains slightly inhibited Sup35 fiber formation in vitro. So far no such inhibitor has been identified. However, Sup35, and specifically [*PSI*⁺] aggregates, were shown to bind the Ssa and Ssb chaperones,[101] and Ssb1 is a known antagonist of de novo prion formation.[47] Also, Rnq1 in the [*PIN*⁺] state binds Sis1,[102,103] and Ssa1 together with its cochaperones, Ydj1 or Sis1, has recently been shown to inhibit Sup35 fiber formation in vitro.[76]

As for the antagonistic prion-prion interactions, it is possible that titration of cellular factors by one type of prion aggregate would compromise the propagation of a heterologous prion.

Cellular Response Model

Schwimmer and Masison[36] proposed this model (Fig. 5) to explain the antagonistic interactions of [*PSI*⁺] and [URE3]. They found that different prions affect the levels of different chaperones in distinct ways and showed a differential sensitivity of different prions to different chaperones. For example they found that the presence of [*PSI*⁺] (and to a lesser extent [URE3]) caused an increase in the expression of Ssa1. Because Ssa1 destabilizes [URE3] but not [*PSI*⁺], selective [URE3] destabilization was attributed to the increase in Ssa1 expression.[36,37] The simultaneous presence of [*PSI*⁺] and [URE3] also resulted in a significant increase in Hsp104 levels, to which [*PSI*⁺] is sensitive and [URE3] is not.[5,7] The latter could explain the reduction in [*PSI*⁺] phenotypic expression. It is also possible, that a chaperone imbalance triggers a more complex cascade of prion-chaperone interactions.[104-106]

A positive effect of a preexisting chaperone on the formation of another prion is also possible within the framework of this model. For example, excess Ssa could additionally facilitate [*PSI*⁺] induction in a [*PIN*⁺] background.[101]

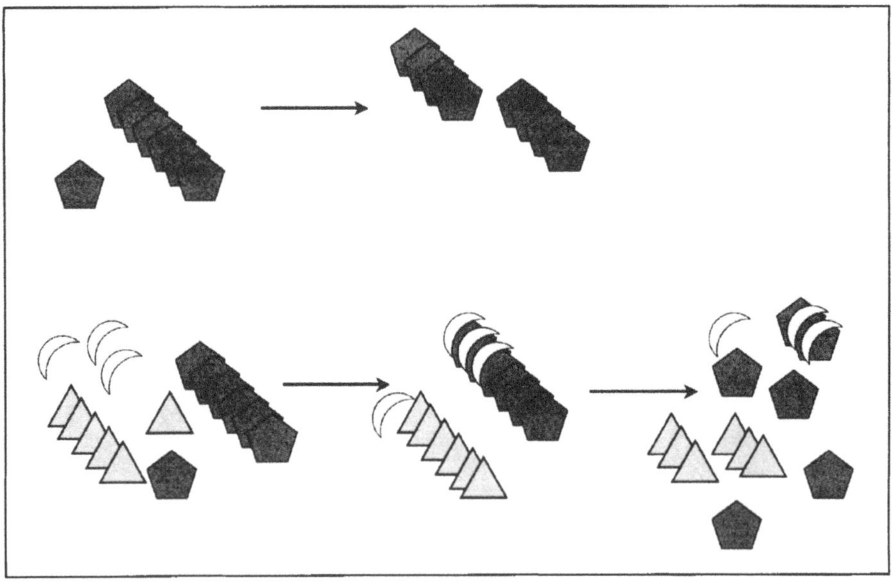

Figure 5. Cellular response model to explain why [*PSI*⁺] destabilizes [URE3]. Top: normal propagation of [URE3] (red pentagons). Bottom: [*PSI*⁺] (linked green triangles) induces the expression of the Ssa1 chaperone (light blur sickles), which destabilizes [URE3] but not [*PSI*⁺]. A color version of this figure can be found at www.eurekah.com.

Since the discovery that the prion domains of several proteins have been retained for hundreds of millions of years,[107] the question of why these ostensibly function-less regions have been conserved has been raised. Evidence suggests that the unstable nature of [*PSI*+] and other prions may provide the organism with an evolutionary advantage, appearing when needed and disappearing when no longer advantageous.[108-110] Indeed, the special advantage prions might offer generally comes with a price tag of a global effect (e.g., on translation termination in the case of Sup35), and the fact that [*PSI*+] and [URE3] are not found in the wild can be interpreted to mean that they are diseases.[56] One possibility is that prions are retained transiently, when their advantages outweigh their negative effects and until more specific mutations are selected for. In this case the possibility of seeding increases the chance of prion formation, and a benign prion, like [*PIN*+], which has been found in nonlaboratory yeast isolates[56] would facilitate this process. Then antagonistic prion-prion interactions would facilitate the loss of prions that do not provide selective advantages.

Acknowledgments

Work in the authors' laboratories was supported by National Science Foundation Grant 0518482 (I.L.D.) and National Institutes of Health Grant GM056350 (S.W.L.). We thank Yakov Vitrenko, Vidhu Mathur, Andrew O'Dell, Michele Kadnar, Catherine Potenski and N. Kaye Horstman for helpful comments on the manuscript.

References

1. Prusiner SB. Novel proteinaceous infectious particles cause scrapie. Science 1982; 216(4542):136-144.
2. Wickner RB. [URE3] as an altered URE2 protein: Evidence for a prion analog in Saccharomyces cerevisiae. Science 1994; 264(5158):566-569.
3. Coustou V, Deleu C, Saupe S et al. The protein product of the het-s heterokaryon incompatibility gene of the fungus Podospora anserina behaves as a prion analog. Proc Natl Acad Sci USA 1997; 94(18):9773-9778.
4. Wickner RB. A new prion controls fungal cell fusion incompatibility. Proc Natl Acad Sci USA 1997; 94(19):10012-10014.
5. Chernoff YO, Lindquist SL, Ono B et al. Role of the chaperone protein Hsp104 in propagation of the yeast prion-like factor [PSI+]. Science 1995; 268(5212):880-884.
6. Derkatch IL, Bradley ME, Zhou P et al. Genetic and environmental factors affecting the de novo appearance of the [PSI+] prion in Saccharomyces cerevisiae. Genetics 1997; 147(2):507-519.
7. Moriyama H, Edskes HK, Wickner RB. [URE3] prion propagation in Saccharomyces cerevisiae: Requirement for chaperone Hsp104 and curing by overexpressed chaperone Ydj1p. Mol Cell Biol 2000; 20(23):8916-8922.
8. Sondheimer N, Lindquist S. Rnq1: An epigenetic modifier of protein function in yeast. Mol Cell 2000; 5(1):163-172.
9. Maddelein ML, Dos Reis S, Duvezin-Caubet S et al. Amyloid aggregates of the HET-s prion protein are infectious. Proc Natl Acad Sci USA 2002; 99(11):7402-7407.
10. King CY, Diaz-Avalos R. Protein-only transmission of three yeast prion strains. Nature 2004; 428(6980):319-323.
11. Tanaka M, Chien P, Naber N et al. Conformational variations in an infectious protein determine prion strain differences. Nature 2004; 428(6980):323-328.
12. Brachmann A, Baxa U, Wickner RB. Prion generation in vitro: Amyloid of Ure2p is infectious. EMBO J 2005; 24(17):3082-3092.
13. Magasanik B. The transduction of the nitrogen regulation signal in Saccharomyces cerevisiae. Proc Natl Acad Sci USA 2005;102(46):16537-16538.
14. Inge-Vechtomov S, Zhouravleva G, Philippe M. Eukaryotic release factors (eRFs) history. Biol Cell 2003; 95(3-4):195-209.
15. Saupe SJ. Molecular genetics of heterokaryon incompatibility in filamentous ascomycetes. Microbiol Mol Biol Rev 2000; 64(3):489-502.
16. Dalstra HJ, van der Zee R, Swart K et al. Non-Mendelian inheritance of the HET-s prion or HET-s prion domains determines the het-S spore killing system in Podospora anserina. Fungal Genet Biol 2005; 42(10):836-847.

17. Steele AD, Emsley JG, Ozdinler PH et al. Prion protein (PrPc) positively regulates neural precursor proliferation during developmental and adult mammalian neurogenesis. Proc Natl Acad Sci USA 2006; 103(9):3416-3421.
18. Zhang CC, Steele AD, Lindquist S et al. Prion protein is expressed on long-term repopulating hematopoietic stem cells and is important for their self-renewal. Proc Natl Acad Sci USA 2006; 103(7):2184-2189.
19. Chien P, Weissman JS, DePace AH. Emerging principles of conformation-based prion inheritance. Annu Rev Biochem 2004; 73:617-656.
20. Bendheim PE, Barry RA, DeArmond SJ et al. Antibodies to a scrapie prion protein. Nature 1984; 310(5976):418-421.
21. Kimura Y, Koitabashi S, Fujita T. Analysis of yeast prion aggregates with amyloid-staining compound in vivo. Cell Struct Funct 2003; 28(3):187-193.
22. Kryndushkin DS, Alexandrov IM, Ter-Avanesyan MD et al. Yeast [PSI+] prion aggregates are formed by small Sup35 polymers fragmented by Hsp104. J Biol Chem 2003; 278(49):49636-49643.
23. Bagriantsev SN, Liebman SW. Specificity of prion assembly in vivo: [PSI+] and [PIN+] form separate structures in yeast. J Biol Chem 2004; 279(49):51042-51048.
24. Baskakov IV. Autocatalytic conversion of recombinant prion proteins displays a species barrier. J Biol Chem 2004; 279(9):7671-7677.
25. Glover JR, Kowal AS, Schirmer EC et al. Self-seeded fibers formed by Sup35, the protein determinant of [PSI+], a heritable prion-like factor of S. cerevisiae. Cell 1997; 89(5):811-819.
26. Taylor KL, Cheng N, Williams RW et al. Prion domain initiation of amyloid formation in vitro from native Ure2p. Science 1999; 283(5406):1339-1343.
27. Thual C, Komar AA, Bousset L et al. Structural characterization of Saccharomyces cerevisiae prion-like protein Ure2. J Biol Chem 1999; 274(19):13666-13674.
28. Schlumpberger M, Wille H, Baldwin MA et al. The prion domain of yeast Ure2p induces autocatalytic formation of amyloid fibers by a recombinant fusion protein. Protein Sci 2000; 9(3):440-451.
29. Dos Reis S, Coulary-Salin B, Forge V et al. The HET-s prion protein of the filamentous fungus Podospora anserina aggregates in vitro into amyloid-like fibrils. J Biol Chem 2002; 277(8):5703-5706.
30. King CY, Tittmann P, Gross H et al. Prion-inducing domain 2-114 of yeast Sup35 protein transforms in vitro into amyloid-like filaments. Proc Natl Acad Sci USA 1997; 94(13):6618-6622.
31. Kayed R, Head E, Thompson JL et al. Common structure of soluble amyloid oligomers implies common mechanism of pathogenesis. Science 2003; 300(5618):486-489.
32. Shorter J, Lindquist S. Hsp104 catalyzes formation and elimination of self-replicating Sup35 prion conformers. Science 2004; 304(5678):1793-1797.
33. Dobson CM. Principles of protein folding, misfolding and aggregation. Semin Cell Dev Biol 2004; 15(1):3-16.
34. Chernoff YO. Amyloidogenic domains, prions and structural inheritance: Rudiments of early life or recent acquisition? Curr Opin Chem Biol 2004; 8(6):665-671.
35. Jones GW, Tuite MF. Chaperoning prions: The cellular machinery for propagating an infectious protein? Bioessays 2005; 27(8):823-832.
36. Schwimmer C, Masison DC. Antagonistic interactions between yeast [PSI+] and [URE3] prions and curing of [URE3] by Hsp70 protein chaperone Ssa1p but not by Ssa2p. Mol Cell Biol 2002; 22(11):3590-3598.
37. Jung G, Jones G, Wegrzyn RD et al. A role for cytosolic hsp70 in yeast [PSI+] prion propagation and [PSI+] as a cellular stress. Genetics 2000; 156(2):559-570.
38. Derkatch IL, Bradley ME, Hong JY et al. Prions affect the appearance of other prions: The story of [PIN+]. Cell 2001; 106(2):171-182.
39. Derkatch IL, Bradley ME, Masse SV et al. Dependence and independence of [PSI+] and [PIN+]: A two-prion system in yeast? EMBO J 2000; 19(9):1942-1952.
40. Wickner RB. Prions of yeast and heat-shock protein 104: 'Coprion' and cure. Trends Microbiol 1995; 3(10):367-369.
41. Aigle M, Lacroute F. Genetical aspects of [URE3], a nonmitochondrial, cytoplasmically inherited mutation in yeast. Mol Gen Genet 1975; 136(4):327-335.
42. Lund PM, Cox BS. Reversion analysis of [psi-] mutations in Saccharomyces cerevisiae. Genet Res 1981; 37(2):173-182.
43. Tikhodeev ON, Gwetmanova EV, Tikhomirova VL et al. Ambiguity of Translation in Yeast. In: Shestakov SV, Tarasov VA, eds. Molecular mechanisms of genetic processes. Moscow: Nauka, 1990:218-228.

44. Chernoff YO, Derkach IL, Inge-Vechtomov SG. Multicopy SUP35 gene induces de-novo appearance of psi-like factors in the yeast Saccharomyces cerevisiae. Curr Genet 1993; 24(3):268-270.
45. Derkatch IL, Chernoff YO, Kushnirov VV et al. Genesis and variability of [PSI] prion factors in Saccharomyces cerevisiae. Genetics 1996; 144(4):1375-1386.
46. Derkatch IL, Bradley ME, Liebman SW. Overexpression of the SUP45 gene encoding a Sup35p-binding protein inhibits the induction of the de novo appearance of the [PSI$^+$] prion. Proc Natl Acad Sci USA 1998; 95(5):2400-2405.
47. Chernoff YO, Newnam GP, Kumar J et al. Evidence for a protein mutator in yeast: Role of the Hsp70-related chaperone ssb in formation, stability, and toxicity of the [PSI] prion. Mol Cell Biol 1999; 19(12):8103-8112.
48. Conde J, Fink GR. A mutant of Saccharomyces cerevisiae defective for nuclear fusion. Proc Natl Acad Sci USA 1976; 73(10):3651-3655.
49. Tuite MF, Mundy CR, Cox BS. Agents that cause a high frequency of genetic change from [PSI$^+$] to [psi$^-$] in Saccharomyces cerevisiae. Genetics 1981; 98(4):691-711.
50. Juliani MH, Da Costa SO. Induction of rho- mutants in Saccharomyces cerevisiae by guanidine hydrochloride. II. Conditions that prevent rho- induction. Mutat Res 1975; 30(3):335-342.
51. Cox BS. Psi, a cytoplasmic suppressor of super-supression in yeasts. Heredity 1965; 20:505-521.
52. Dagkesamanskaya AR, Ter-Avanesyan MD. Interaction of the yeast omnipotent suppressors SUP1 (SUP45) and SUP2 (SUP35) with non-Mendelian factors. Genetics 1991; 128(3):513-520.
53. Chernoff YO, Inge-Vechtomov SG, Derkach IL et al. Dosage-dependent translational suppression in yeast Saccharomyces cerevisiae. Yeast 1992; 8(7):489-499.
54. Ganusova EE, Ozolins LN, Bhagat S et al. Modulation of prion formation, aggregation, and toxicity by the actin cytoskeleton in yeast. Mol Cell Biol 2006; 26(2):617-629.
55. Resende CG, Outeiro TF, Sands L et al. Prion protein gene polymorphisms in Saccharomyces cerevisiae. Mol Microbiol 2003; 49(4):1005-1017.
56. Nakayashiki T, Kurtzman CP, Edskes HK et al. Yeast prions [URE3] and [PSI$^+$] are diseases. Proc Natl Acad Sci USA 2005; 102(30):10575-10580.
57. Osherovich LZ, Weissman JS. Multiple Gln/Asn-rich prion domains confer susceptibility to induction of the yeast [PSI$^+$] prion. Cell 2001; 106(2):183-194.
58. Bradley ME, Edskes HK, Hong JY et al. Interactions among prions and prion "strains" in yeast. Proc Natl Acad Sci USA 2002; 99(Suppl 4):16392-16399.
59. Fraser H, Dickinson AG. The sequential development of the brain lesion of scrapie in three strains of mice. J Comp Pathol 1968; 78(3):301-311.
60. Schlumpberger M, Prusiner SB, Herskowitz I. Induction of distinct [URE3] yeast prion strains. Mol Cell Biol 2001; 21(20):7035-7046.
61. Krishnan R, Lindquist SL. Structural insights into a yeast prion illuminate nucleation and strain diversity. Nature 2005; 435(7043):765-772.
62. Tanaka M, Chien P, Yonekura K et al. Mechanism of cross-species prion transmission: An infectious conformation compatible with two highly divergent yeast prion proteins. Cell 2005; 121(1):49-62.
63. Nelson R, Sawaya MR, Balbirnie M et al. Structure of the cross-beta spine of amyloid-like fibrils. Nature 2005; 435(7043):773-778.
64. Liebman SW. Structural clues to prion mysteries. Nat Struct Mol Biol 2005; 12(7):567-568.
65. Zhou P, Derkatch IL, Uptain SM et al. The yeast non-Mendelian factor [ETA$^+$] is a variant of [PSI$^+$], a prion-like form of release factor eRF3. EMBO J 1999; 18(5):1182-1191.
66. Liebman SW, Bagriantsev SN, Derkatch IL. Biochemical and genetic methods for characterization of [PIN$^+$] prions in yeast. Methods 2006; 39(1):23-34.
67. Derkatch IL, Uptain SM, Outeiro TF et al. Effects of Q/N-rich, polyQ, and non-poly-Q amyloids on the de novo formation of the [PSI$^+$] prion in yeast and aggregation of Sup35 in vitro. Proc Natl Acad Sci USA 2004; 101(35):12934-12939.
68. Meriin AB, Zhang X, He X et al. Huntington toxicity in yeast model depends on polyglutamine aggregation mediated by a prion-like protein Rnq1. J Cell Biol 2002; 157(6):997-1004.
69. Gokhale KC, Newnam GP, Sherman MY et al. Modulation of prion-dependent polyglutamine aggregation and toxicity by chaperone proteins in the yeast model. J Biol Chem 2005; 280(24):22809-22818.
70. Perutz MF. Glutamine repeats and neurodegenerative diseases: Molecular aspects. Trends Biochem Sci 1999; 24(2):58-63.
71. Perutz MF, Pope BJ, Owen D et al. Aggregation of proteins with expanded glutamine and alanine repeats of the glutamine-rich and asparagine-rich domains of Sup35 and of the amyloid beta-peptide of amyloid plaques. Proc Natl Acad Sci USA 2002; 99(8):5596-5600.

72. Legname G, Baskakov IV, Nguyen HO et al. Synthetic mammalian prions. Science 2004; 305(5684):673-676.
73. Collin P, Beauregard PB, Elagoz A et al. A nonchromosomal factor allows viability of Schizosaccharomyces pombe lacking the essential chaperone calnexin. J Cell Sci 2004; 117(Pt 6):907-918.
74. DebBurman SK, Raymond GJ, Caughey B et al. Chaperone-supervised conversion of prion protein to its protease-resistant form. Proc Natl Acad Sci USA 1997; 94(25):13938-13943.
75. Inoue Y, Taguchi H, Kishimoto A et al. Hsp104 binds to yeast sup35 prion fiber but needs other factor(s) to sever it. J Biol Chem 2004; 279(50):52319-52323.
76. Krzewska J, Melki R. Molecular chaperones and the assembly of the prion Sup35p, an in vitro study. EMBO J 2006; 25(4):822-833.
77. Uptain SM, Sawicki GJ, Caughey B et al. Strains of [PSI$^+$] are distinguished by their efficiencies of prion-mediated conformational conversion. EMBO J 2001; 20(22):6236-6245.
78. Santoso A, Chien P, Osherovich LZ et al. Molecular basis of a yeast prion species barrier. Cell 2000; 100(2):277-288.
79. DePace AH, Weissman JS. Origins and kinetic consequences of diversity in Sup35 yeast prion fibers. Nat Struct Biol 2002; 9(5):389-396.
80. Chien P, Weissman JS. Conformational diversity in a yeast prion dictates its seeding specificity. Nature 2001; 410(6825):223-227.
81. Hara H, Nakayashiki T, Crist CG et al. Prion domain interaction responsible for species discrimination in yeast [PSI$^+$] transmission. Genes Cells 2003; 8(12):925-939.
82. Vanik DL, Surewicz KA, Surewicz WK. Molecular basis of barriers for interspecies transmissibility of mammalian prions. Mol Cell 2004; 14(1):139-145.
83. Bradley ME, Liebman SW. Destabilizing interactions among [PSI$^+$] and [PIN$^+$] yeast prion variants.Genetics 2003; 165(4):1675-1685.
84. Satpute-Krishnan P, Serio TR. Prion protein remodelling confers an immediate phenotypic switch. Nature 2005; 437(7056):262-265.
85. Salnikova AB, Kryndushkin DS, Smirnov VN et al. Nonsense suppression in yeast cells overproducing Sup35 (eRF3) is caused by its nonheritable amyloids. J Biol Chem 2005; 280(10):8808-8812.
86. Perutz MF, Johnson T, Suzuki M et al. Glutamine repeats as polar zippers: Their possible role in inherited neurodegenerative diseases. Proc Natl Acad Sci USA 1994; 91(12):5355-5358.
87. Petkova AT, Ishii Y, Balbach JJ et al. A structural model for Alzheimer's beta -amyloid fibrils based on experimental constraints from solid state NMR. Proc Natl Acad Sci USA 2002; 99(26):16742-16747.
88. Lipfert J, Franklin J, Wu F et al. Protein misfolding and amyloid formation for the peptide GNNQQNY from yeast prion protein Sup35: Simulation by reaction path annealing. J Mol Biol 2005; 349(3):648-658.
89. DePace AH, Santoso A, Hillner P et al. A critical role for amino-terminal glutamine/asparagine repeats in the formation and propagation of a yeast prion. Cell 1998; 93(7):1241-1252.
90. Niewold TA, Hol PR, van Andel AC et al. Enhancement of amyloid induction by amyloid fibril fragments in hamster. Lab Invest 1987; 56(5):544-549.
91. Lundmark K, Westermark GT, Nystrom S et al. Transmissibility of systemic amyloidosis by a prion-like mechanism. Proc Natl Acad Sci USA 2002; 99(10):6979-6984.
92. Ganowiak K, Hultman P, Engstrom U et al. Fibrils from synthetic amyloid-related peptides enhance development of experimental AA-amyloidosis in mice. Biochem Biophys Res Commun 1994; 199(1):306-312.
93. Johan K, Westermark G, Engstrom U et al. Acceleration of amyloid protein A amyloidosis by amyloid-like synthetic fibrils. Proc Natl Acad Sci USA 1998; 95(5):2558-2563.
94. Lundmark K, Westermark GT, Olsen A et al. Protein fibrils in nature can enhance amyloid protein A amyloidosis in mice: Cross-seeding as a disease mechanism. Proc Natl Acad Sci USA 2005; 102(17):6098-6102.
95. Kisilevsky R, Lemieux L, Boudreau L et al. New clothes for amyloid enhancing factor (AEF): Silk as AEF. Amyloid 1999; 6(2):98-106.
96. O'Nuallain B, Williams AD, Westermark P et al. Seeding specificity in amyloid growth induced by heterologous fibrils. J Biol Chem 2004; 279(17):17490-17499.
97. Edskes HK, Gray VT, Wickner RB. The [URE3] prion is an aggregated form of Ure2p that can be cured by overexpression of Ure2p fragments. Proc Natl Acad Sci USA 1999; 96(4):1498-1503.
98. Edskes HK, Wickner RB. Conservation of a portion of the S. cerevisiae Ure2p prion domain that interacts with the full-length protein. Proc Natl Acad Sci USA 2002; 99(Suppl 4):16384-16391.
99. Osherovich LZ, Weissman JS. The utility of prions. Dev Cell 2002; 2(2):143-151.

100. Zhou P, Derkatch IL, Liebman SW. The relationship between visible intracellular aggregates that appear after overexpression of Sup35 and the yeast prion-like elements [PSI⁺] and [PIN⁺]. Mol Microbiol 2001; 39(1):37-46.
101. Allen KD, Wegrzyn RD, Chernova TA et al. Hsp70 chaperones as modulators of prion life cycle: Novel effects of Ssa and Ssb on the Saccharomyces cerevisiae prion [PSI⁺]. Genetics 2005; 169(3):1227-1242.
102. Sondheimer N, Lopez N, Craig EA et al. The role of Sis1 in the maintenance of the [RNQ⁺] prion. EMBO J 2001; 20(10):2435-2442.
103. Lopez N, Aron R, Craig EA. Specificity of class II Hsp40 Sis1 in maintenance of yeast prion [RNQ⁺]. Mol Biol Cell 2003; 14(3):1172-1181.
104. Song Y, Masison DC. Independent regulation of Hsp70 and Hsp90 chaperones by Hsp70/Hsp90-organizing protein Sti1 (Hop1). J Biol Chem 2005; 280(40):34178-34185.
105. Song Y, Wu YX, Jung G et al. Role for Hsp70 chaperone in Saccharomyces cerevisiae prion seed replication. Eukaryot Cell 2005; 4(2):289-297.
106. Jones G, Song Y, Chung S et al. Propagation of Saccharomyces cerevisiae [PSI⁺] prion is impaired by factors that regulate Hsp70 substrate binding. Mol Cell Biol 2004; 24(9):3928-3937.
107. Nakayashiki T, Ebihara K, Bannai H et al. Yeast [PSI⁺] "prions" that are crosstransmissible and susceptible beyond a species barrier through a quasi-prion state. Mol Cell 2001; 7(6):1121-1130.
108. Eaglestone SS, Cox BS, Tuite MF. Translation termination efficiency can be regulated in Saccharomyces cerevisiae by environmental stress through a prion-mediated mechanism. EMBO J 1999; 18(7):1974-1981.
109. True HL, Berlin I, Lindquist SL. Epigenetic regulation of translation reveals hidden genetic varn provides a mechanism for genetic variation and phenotypic diveiation to produce complex traits. Nature 2004; 431(7005):184-187.
110. True HL, Lindquist SL. A yeast prion provides a mechanism for genetic variation and phenotypic diversity. Nature 2000; 407(6803):477-483.

CHAPTER 5

Prion Stability

Brian S. Cox, Lee Byrne and Mick F. Tuite*

Abstract

The rate of spontaneous change from ψ^- to the ψ^+ condition determined in yeast by states of the Sup35p protein is briefly discussed, together with the conditions necessary for such change to occur. Conditions that promote and which affect the rate of induction of ψ^+ in Sup35p and of other prion-forming proteins to their respective prion forms are also discussed. These include the influence of the amount of nonprion protein, the presence of other prions, the activity of chaperones, and brief descriptions of the role of native sequences in the proteins and how alteration of sequences in prion-forming proteins influence the rate of induction of [prion$^+$] and amyloid forms.

The second part of this article discusses the conditions which affect the reversion of ψ^+ to ψ^-, including factors which affect the copy-number of prion "seeds" or propagons and their partition. The principal factor discussed is the activity of the chaperone Hsp104, but the existence of other factors, such a protein sequence and of other, less well-studied agents, is touched upon and comparisons are made, as appropriate, with studies with other yeast prions.

We conclude with a discussion of models of maintenance, in particular that of Tanaka et al recently published in *Nature* (2006),[6] which provides much insight into the phenotypic and genetic parameters of the numerous "variants" of prions increasingly being described in the literature.

Introduction

It is a commonplace observation that the prion forms of proteins are stable in inheritance; indeed it is implicit in the definition of a prion and distinguishes them from such forms as conformers, aggregates, polymers and amyloids which are neither infectious nor heritable. In this context, stability implies more than mere chemical stability and more than just synthesis, since the alternative nonprion state is equally stable. This implies self-reproduction, that is a requirement for a preexisting template. The relevance of fungal prions to the mammalian variety which cause infectious diseases arises through the commonality of infection and heredity, which has been remarked upon by many distinguished biologists. One nice example of this commonality was pointed out by Francois Jacob, talking about the temperate bacteriophage phage λ, which may be observed either as a fatal disease of its host *E. coli*, or as one of its genes with a unique location, conferring resistance and lysogeny. The difference lies in the particular mode of reproduction and transmission of the alternative states. In the world of prions, both properties, of infection and heritability, are illustrated by het-s, a prion of the fungus *Podospora anserina.*[1] When two mycelia of this fungus fuse, the [Het-s] prion will migrate from one to the other and spread through its new host mycelium. That is infection: but the prion may also be inherited through the spores produced by meiosis in the asci. In each case, the need and, with reservations, sufficiency of a template is characteristic of the genetic nature of the phenomena.

*Corresponding Author: Mick F. Tuite—Department of Biosciences, University of Kent, Canterbury, Kent CT2 7NJ, U.K. Email: m.f.tuite@kent.ac.uk

Protein-Based Inheritance, edited by Yury O. Chernoff. ©2007 Landes Bioscience.

There are three established native prions of the yeast *Saccharomyces cerevisiae* whose genes are known: ψ (*SUP35*), [URE3] (*URE2*) and [*PIN*] (*RNQ1*).[2-5,14] There is also a gene, *NEW1*, coding for a protein of unknown function, which has sequences which promote prion-like behaviour in gene fusions but the native protein it codes for has not been shown to occur in a prion form.[5]

There are additionally numerous other synthetic prions, consisting of sequences derived from those of prion-forming proteins modified by deletion, mutagenesis, or by fusion with heterologous natural or artificial sequences for functional modification or as reporters. Of the native prions of yeast, ψ has, perhaps because of its relative stability and easily scored phenotype, lent itself to studies of factors affecting reproductive stability. This article will summarize some of these studies and we will review studies on the cellular events and variables that might affect its stability. We will also consider the kinetic constraints underlying stable transmission of alternative [prion$^+$] or [prion$^-$] states, which have recently been elegantly analysed by Tanaka et al.[6]

Yeast cells may exist in either of two states of considerable stability—ψ^+ and ψ^-. Either may revert to the other spontaneously, but in normal circumstances they do so very rarely.

Changes from ψ^- to ψ^+

Spontaneous Reversion

The ψ^- state is stable in growing cultures, in stored cultures under a variety of conditions and in the course of sporulation. Few measurements are available for reversion to ψ^+ in any of these conditions. In one study, a less than rigorous fluctuation test suggested a rate of ~1 x 10^{-7} celldivision^{-1}.[7] An earlier study in which ψ^- states had been induced by various treatments of ψ^+ strains showed that many, but not all of these ψ^- revertant strains could change spontaneously to ψ^+.[8] The treatments causing the initial ψ^+ to ψ^- reversion included growth in the presence of methanol, KCl, DMSO or guanidine hydrochloride and treatment with the conventional mutagens EMS, nitrosoguanidine and UV. At least some of the ψ^- strains from all of the treatments, with one exception, could change spontaneously from ψ^- back to ψ^+. The frequencies observed ranged from 2.8 x 10^{-3} to 8.5 x10^{-8}, with a median frequency of ~6 x 10^{-6}.[a] The exceptions were 18 ψ^- revertants induced by 5 mM guanidine HCl, none of which yielded ψ^+ among 10^8 cells challenged from each.[8]

These experiments were performed under the paradigm of the time, namely that genetic determinants were nucleic acid. It was deduced that guanidine HCl caused deletions in the ψ determinant as it does in mitochondrial DNA.[9] It has since become clear that ψ^- is a hyperstable state and simply does not convert to ψ^+ unless another prion ([*PIN*$^+$]) is present[b].[10] Although [*PIN*$^+$] and ψ^+ are lost more or less independently when Hsp104 activity is inhibited,[18] under most curing protocols, including that used by Lund and Cox,[8] guanidine HCl almost invariably cures both ψ^+ and [*PIN*$^+$] together. It follows that all experiments on the spontaneous or induced conversion of the ψ^- state to ψ^+ are conditional upon and must take into account the implications of the presence of another prion.

Dependence of spontaneous conversion on the presence of another kind of prion is not a property of all prions. The prion-negative state of [URE3], for example is much more labile than that of ψ[2,11] and is not absolutely dependent on the strain being [*PIN*$^+$].[12] This will be discussed at greater length below, but more particularly in Chapter 4 by Liebman and Derkach in this volume.[51]

---

[a] These frequencies do not necessarily represent different inherent instabilities, and are more likely to be technical artefacts

[b] As usual there is an exception.[18] ψ^- can convert to ψ^+ in a [*pin*$^-$] background when a truncated version of *SUP35* with a small extension picked up from its vector is over-expressed.

Induction

The rate at which ψ⁻ converts to ψ⁺ can be affected by various factors. These include:

- Amount of normal protein
- Presence of other prions (the [*PIN*] effect)
- Activity of chaperones
- DNA sequences in the prion-forming domain and outside it
- Acquisition of a prion form of Sup35p

The Influence of the Amounts of Protein

The native yeast prion proteins, ψ, and [URE3], change from the [prion⁻] to the [prion⁺] state more frequently if the parent genes are over-expressed.[2,11,13] The native Rnq1p and New1p proteins have not been assayed for this effect, but it is observed when their prion-forming domains (PFDs) are fused to an indicator sequence, such as the C-terminal domain of Sup35p.[5,14] Both ψ and [URE3] show about a 100-fold increase in the frequency of change from [prion⁻] to [prion⁺] states. The effect is even more pronounced when truncated portions of the genes, containing the PFDs are over-expressed.[15,16] In these cases conversion to the prion form is marked by the coupled conversion of the native (full-length) protein.

This property was proposed by Wickner as one of the indicators that a heritable non-Mendelian state was due to a prion-based determinant and is now regarded as a fundamental property of such systems (e.g., Roberts and Wickner[17]). This property has been the basis of searches for other prions in yeast, using high-expression gene libraries.[4]

The rationale behind this argument is that higher concentrations of the native protein enhance the probability that a spontaneous conversion will occur in one or more of the nonprion-form molecules present and that this will trigger the seeded conversion of the remainder. While this turns out to be what is observed, wherever it is tested, matters are not as simple as that, as we discuss in the next section.

The Influence of Other Prions: ψ and [*PIN*]

We have noted above that when ψ⁺ is cured by allowing cell division in the presence of guanidine HCl, it does not seem to be revertible to ψ⁺. With the exception noted above (footnote, p.2), this has turned out to be true even when the *SUP35* gene is over-expressed and also when only the N or NM domains or other potent ψ⁺-inducing constructs are over-expressed. This is, of course, an anomaly, incompatible with one of the important criteria defining a prion proposed by Wickner namely that, as long as the gene coding for the prion-forming protein is present and active, the [prion⁻] state should always be convertible to the [prion⁺].

Liebman and her coworkers found that the de novo conversion or induction of ψ⁺ by over-expression required the presence of another factor. This they named it [*PIN*⁺] and showed that it too was inherited in a non-Mendelian fashion.[10] In due course, they identified [*PIN*⁺] as the prion of Rnq1p, [*RNQ*⁺].[4] In the same paper, they showed that [URE3⁺] and the over-expressed products of ten other genes could also function as [*PIN*⁺].

It also seems to be the case that a heritable form of the heterologous prion may not be necessary for this interaction: amyloid-like aggregates may be sufficient. Firstly, Osherovich and Weissman[5] constructed fusions of the Asn-Gln-rich sequences from the prion-forming domains of *NEW1* and *RNQ1*, in each case with GFP. When either of these was coover-expressed with *SUP35PFD-GFP* in [*pin*⁻] strains, both GFP aggregates, visible microscopically, and heritable ψ⁺ convertants were obtained, but they remained [*pin*⁻].[5] The rate of induction was comparable to that found when the strains were heritably [*PIN*⁺], about 6% when the *NEW1* PFD sequences were over-expressed and one tenth of that in over-expression of the *RNQ1* PFD sequences.

Secondly, Derkach et al[4] have found that over-expression of poly-Q tracts, up to three times longer than those in huntingtin genes associated with Huntington's Disease, allow the induction of ψ⁺ in a [*pin*⁻] background. The longer tracts form amyloid readily in yeast cells, and in cells with amyloid from these longer tracts, ψ^+ may be induced by over-expression of the NM domains of *SUP35*.[19]

[*PIN*$^+$] is not only necessary for prion formation by Sup35p, but is also necessary for amyloid formation when Sup35p is over-expressed. Most of this amyloid appears not to be heritable: 90% of the cells remain [ψ^-], although they allow read-through of nonsense codons.[20] It would seem that the formation of amyloid is not sufficient to make a prion: either amyloid is a precursor and a further change in protein conformation must occur to convert it to prion or amyloid and prion are all part of a range of conformers most of which present as amyloid, but with some being susceptible to fragmentation by Hsp104 and being therefore prions. It has been noted that the new "spontaneously-induced" ψ^+ strains display a wide range of 'variant' types.[4,5,15,18-20] It may be that in a cell over-expressing Sup35p, provided some [*PIN*$^+$]-like function exists, a wide variety of refolded type of molecule is produced. If these possess various susceptibilities to the action of Hsp104, then each would present a 'variant' phenotype (see Tanaka et al,[6] and discussion below). The ψ^+ variant dependent on elevated levels of Hsp104 described by Borchsenius et al[58] is an example of a conformer intermediate between those qualifying as amyloid or as prion.

It is not known what the cause is of these effects. Current speculations lean to the idea that amyloid polymers assist each others' condensation even if they have not enough structural similarities to form copolymers. What does seem to be the case is that the conformational change needed to form self-replicating prions, at least of Sup35p, is, if not wholly impossible in vivo, at least extremely rare in the absence of some amyloid cofactor. In living cells, the normal conformer of Sup35p is hyperstable.

In summary, the presence of any of a wide range of heterologous amyloid proteins, either in prion form or otherwise, seems to be a necessary precondition for the formation of amyloid from Sup35 protein. Conversion of this amyloid to prion is relatively infrequent, clearly showing that a further, or perhaps a completely different kind of conformational change is required to make the Sup35p amyloid heritable. A similar situation may apply to Rnq1 protein, at least in the hybrid RNQ-GFP version, but the absolute requirement for heterologous amyloid probably has not been so rigorously tested. Certainly, ψ^+ prions greatly facilitate its conversion. Ure2 protein responds in a similar way to heterologous prions, except that ψ^+ has an antagonistic effect on its conversion. However, the presence of another prion ([*PIN*$^+$]) is not an essential condition for reversion from [ure3-0] to [URE3$^+$].

Sequence

The sequences defining prion behaviour of proteins have been the subject of much analysis (see Chapter 2 in this volume by Tuite and Cox ref. 52). The question we address here is whether any specifically affect the frequency of spontaneous reversion from the [prion$^-$] to the [prion$^+$] state. One question is whether any exist which determine a permanent [prion$^+$] condition. The criterion for this would be that the inheritance of the prion condition would be non-Mendelian, but there would be a Mendelian segregation for curability.[c] A less extreme situation would be that the spontaneous rate of prion conversion would simply be very much higher in proteins with one sequence than another. There are surprisingly few examples of this. It seems to be true that the nonprion-forming domains, M and C, of Sup35p inhibit the rate of formation of ψ^+ de novo, at least in over-expression experiments.[10,21] These experiments served

[c] One of the characteristics so far observed of prions in fungi is that all authentic native prions are dependent on Hsp104 activity for reproduction. Inactivating or eliminating Hsp104 leads to their being "cured". A sequence which automatically forms "prion" would not show curing. The only indication that the phenotype was due to a prion form of a protein would be if it were to convert an alternative, different, sequence of the protein to a prion form, dependent on Hsp104. This emphasises the requirement for ***seeding*** as a property of a prion protein to a critical role in defining the phenomenon.

to identify the prion-forming domains of this protein. Similarly, specific fragments of Ure2p yield [URE3$^+$] on over-expression.[11] The differences can be quite dramatic: when full length Ure2p is over-expressed, Masison et al[11] found 1.1 x 10^{-4} [URE3$^+$] revertants, but the numbers were 1.7 x 10^{-3} and 1.758 x 10^{-2} for over-expression of a residue 1-65 and a residue 1-80 fragment respectively.

Fernandez-Bellot et al, found a mutated allele of Ure2 which elevated spontaneous rates of reversion from [ure3-o] to [URE3$^+$] some 1000-fold.[22] The mutant gene turned out to have 14 point mutations, two of them in the PFD. These two alone did not much increase the frequency of reversion above that found in the wild-type, but adding one of the mutations found in the functional domain raised the frequency 500x. Such cis-interactions implicate non-PFD domains in prion formation and stability and echo the influence of polymorphisms in PrP on susceptibility to CJD in humans. It is interesting, given this result, that mistranslation per se may elevate spontaneous prion-forming rates in Ure2p.[23] They compared the effects of a general mistranslation drug, G418, with cycloheximide. In similar experiments with ψ^- reversion, Koloteva-Levin et al found no effect with a drug which targets proline specifically.[24]

Naturally, much attention has been given to the details of the PFD sequences and their role in [prion$^-$] to [prion$^+$] conversion. The first of these was by Liu and Lindquist, who set out to mimic the effect noted in humans that expanded numbers of the oligopeptide repeats found at positions 50 - 94 in PrP render subjects more prone to developing CJD.[25] Liu and Lindquist found that two extra copies of one of the analogous repeats in the Sup35p sequence led to a 5000x increase in the spontaneous frequency of ψ^+ formation.

This was complemented by Parham et al who one at a time removed repeats and found that first replication and then inclusion in aggregates were affected as more and more repeats were removed.[26]

This was followed by a random PCR-mutagenesis study by De Pace et al of the N-terminal prion-forming domain of Sup35.[27] They screened the mutant libraries by transforming a ψ^+ strain and looking for transformants which lost the ability to suppress the *ade1-14* nonsense mutation. Failure to suppress is a signal of the presence of significant amounts of soluble, active Sup35p (eRF3p). They classified these into two categories—those which failed to suppress because they prevented replication of the native ψ^+ prion (PNM) and those which did so because they failed to form or be recruited into wild-type amyloid in the first place (ASU). All the mutants they picked up fell into the region coding for the N-terminal 33 residues, which covers the Q/N-rich region of the protein. Rather than test each mutant for the ability to form prion spontaneously de novo, they replaced the whole stretch with poly-Q, and found that indeed over-expression of such constructs induced the ψ^+ state.[d] This effectively identifies the Q/N-rich region as critical for prion formation and propagation and Osherovich and Weissman subsequently showed that almost any poly-Q/N-rich sequence in this region supports de novo reversion.[5]

Many other yeast Sup35 proteins have PFDs very similar, but not identical to that of *S. cerevisiae*. When these heterologous sequences are substituted for the bases coding for first 39 residues of *S. cerevisiae* Sup35p, although they themselves can form stable prions on over-expression, they cannot induce prion formation in native, resident Sup35p unless it too has the same sequence.[28] There is thus a firm species barrier to the transmission of the prion form and it resides in the residues responsible for initiating aggregation, as defined by De Pace et al.[27] It is beginning to look as if there are two functional domains in the PFD, one affecting the initial formation of amyloid and the other the replication necessary to make keep amyloid heritable. A domain-swapping analysis by Osherovich et al seems to confirm this idea,[29] as does the study by Borchsenius et al in which a deletion which removes part of the oligopeptide

[d] All these experiments were done in the presence of the wild-type *SUP35* gene, so whether the effects are dependent on its presence or not is not clear.

repeat (OR) region reduces the ability of the protein to propagate stably, without greatly affecting its ability to aggregate or induce ψ⁺ formation.[55] Furthermore, a recent paper by Crist et al, describes experiments in which oligopeptide repeats from different yeast were substituted for the native *S. cerevisiae* repeats.[30] They show that derivative Sup35p formed aggregates and inactivated Sup35p function just as the native protein does, i.e., mimicked ψ⁺. The pseudo-ψ⁺ state could be induced by native ψ⁺, but these prions did not require Hsp104 activity for replication. In a negative way, this is consistent with the idea that the target for Hsp104 replication of ψ is the oligopeptide repeat region in Sup35. It was also observed that the pseudo-ψ⁺, which the authors call [*PHI*⁺] could arise spontaneously at 1000x the rate of ψ⁺, and without benefit of [*PIN*⁺].

Acquisition of a Prion Form of Sup35p

In addition to de novo conversion, ψ⁺ prion forms may be acquired by mating, inherited through cell division or by transfection.[31,32] Tanaka et al[6] and see below, have also provided evidence that as little as a single molecule of the prion form is sufficient to convert or start the rapid conversion of a substantial majority of soluble Sup35p prion form in vitro. For just how rapidly this can happen in vivo, see Satpute-Krishnan and Serio.[33] When cells start with only one or very few prion molecules, conversion is, as expected, Malthusian and occurs with a doubling time of eighteen minutes.[6,7,42] It is exactly this property that provides the switch between the two stable states of prion-forming proteins and lies at the basis of the stability of [prion⁺] forms which we discuss next.

Changes from ψ⁺ to ψ⁻

Spontaneous Loss

ψ⁺ is very stable, sometimes almost as stable as ψ⁻. ψ was first discovered as a ψ⁻ mutation in a handful of tetrads which had been expected, being homozygous for a weak super-suppressor, *SUQ5* suppressing the red colour of an *ade2-1* mutant, all to be white. Instead, three or four of the segregants had red sectors. The red reversions failed to segregate when crossed back to white *SUQ5 ade2-1* parents. With one notable exception, neither I nor my colleagues have seen anything like this in our strains since, four decades and thousands of tetrads later. ψ⁺ is equally stable in mitosis: once again it is very uncommon to see a red sector or colony on plates growing colonies, at least of strong ψ⁺ strains.

Nevertheless they occur. The rate of spontaneous loss of ψ⁺ has never been adequately quantified although good systems exist for selecting for the change in phenotype.

Observation of great stability in inheritance raises the question of how it is achieved. There are three situations which might affect it. One is the copy number of a putative determinant. Another is a system for partitioning a low copy-number determinant, as exists for chromosomes, low-copy number bacterial plasmids and, indeed for the relatively high copy-number yeast 2μ plasmid.[34] A third possibility is a feed-back regulatory system which switches on a pathway in response to changes in quantity of a significant component. The yeast 2μ has a feed-back system regulating its amplification, for example.[56] None of these systems is mutually exclusive and underlying all of them is a requirement to increase the numbers of the determinants at the same rate as the cells multiply.

Curing Studies and the Origins of Genetic Stability

The discovery that the inheritance of yeast prions is entirely (but not exclusively) dependent on the activity of the chaperone heat shock protein, Hsp104[35] has made possible a number of studies on the factors maintaining prion stability.

The Kinetics of Curing with Guanidine Hydrochloride and the Quantification of Prion Molecules

Sometime before 1981, Tuite discovered that mM quantities of the denaturing agent guanidine hydrochloride "cured" ψ^+ cells to ψ^- with 100% efficiency.[36,37] It has since been suggested that because such concentrations of guanidine HCl also inhibit Hsp104 activity in vitro and in vivo that ψ^+ curing results also from this inhibition.[38,39,40] The curing occurs only in actively growing cultures. It is not immediate upon addition of the guanidine HCl, but shows a considerable lag before ψ^- cells begin to appear.[41] There are three features of the kinetics of this process that are worth notice. Firstly, once curing starts to be observable, the decline in the proportion of ψ^+ cells is exponential and halves at each cell generation. In fact, the entire curve, lag and all, can be fitted with models assuming a halving of average propagon numbers with each generation, a Poisson distribution of "seeds" among cells and that a single "seed" or propagon is sufficient to render a cell genetically ψ^+, as recently demonstrated by infection experiments,[31] and by further experimental and theoretical development of the Eaglestone model[7,42,43] (Fig. 1A).

Secondly, propagons do not appear to be destroyed during growth in guanidine HCl. The number of genetically ψ^+ cells in the population reaches a maximum asymptotically during the "curing" phase and thereafter remains constant indefinitely;[44] (Fig. 1B). We suppose this number represents ψ^+ cells that are still dividing, but unable to segregate more than one prion molecule at a time, thereby giving rise to one ψ^- and one ψ^+ daughter at each division.

The "copy-number" of the ψ^+ prion has been estimated making an assumption that there is some particulate determinant acting as a "seed", i.e., a propagon. Numbers of these propagons are supposed to be distributed at random between mother and daughter cells at cell division. Eaglestone et al[41] suggested that the reason for the kinetics of guanidine-promoted curing was that Hsp104 was required for the replication or division of seeds in a cycle involving the accretion of soluble Sup35p molecules to a seed, their consequent conversion to the prion form followed by the Hsp104-mediated fragmentation, as suggested by Kushnirov and Ter-Avanesyan in 1998.[45] The lag observed when curing is effected by growth in guanidine HCl,[41] by competition through over-expression of an Hsp104 double mutant[39] or by loss of the *Hsp104* gene,[46] would be a segregation lag dependent on the dilution of the seeds by a half in every cell generation. A simple model was developed whereby the length of the lag could be used to estimate the average copy-number of propagons.[41]

The implications of this model have been tested both by growth kinetic experiments and at the molecular level,[7,42] and more refined mathematical models have been developed to arrive at estimates of propagon (seed) number.[43] These numbers have turned out to be very variable and often strain-dependent but they are always quite high, ranging from 100 to over 1,000 per cell. Similar experiments with [URE3] suggest the copy number of that, less stable prion, is about 20.[47]

ψ^--*Inducing Agents*

The stability of the ψ^+ prion has also been found to be affected by growth in a variety of stress conditions.[37,48] However, the stress most commonly applied in studies of this organism, namely heat shock, has no effect on ψ stability.

On the other hand Tuite et al also found that reversion to ψ^- was induced by conventional mutagens such as UV and nitrosoguanidine or EMS and was subject to DNA repair mechanisms, including photoreactivation of UV damage and excision repair.[49] Since we must now accept that the determinant of ψ is not DNA, it would seem that there are, nevertheless, uninherited, forms of DNA damage which affect its stability.

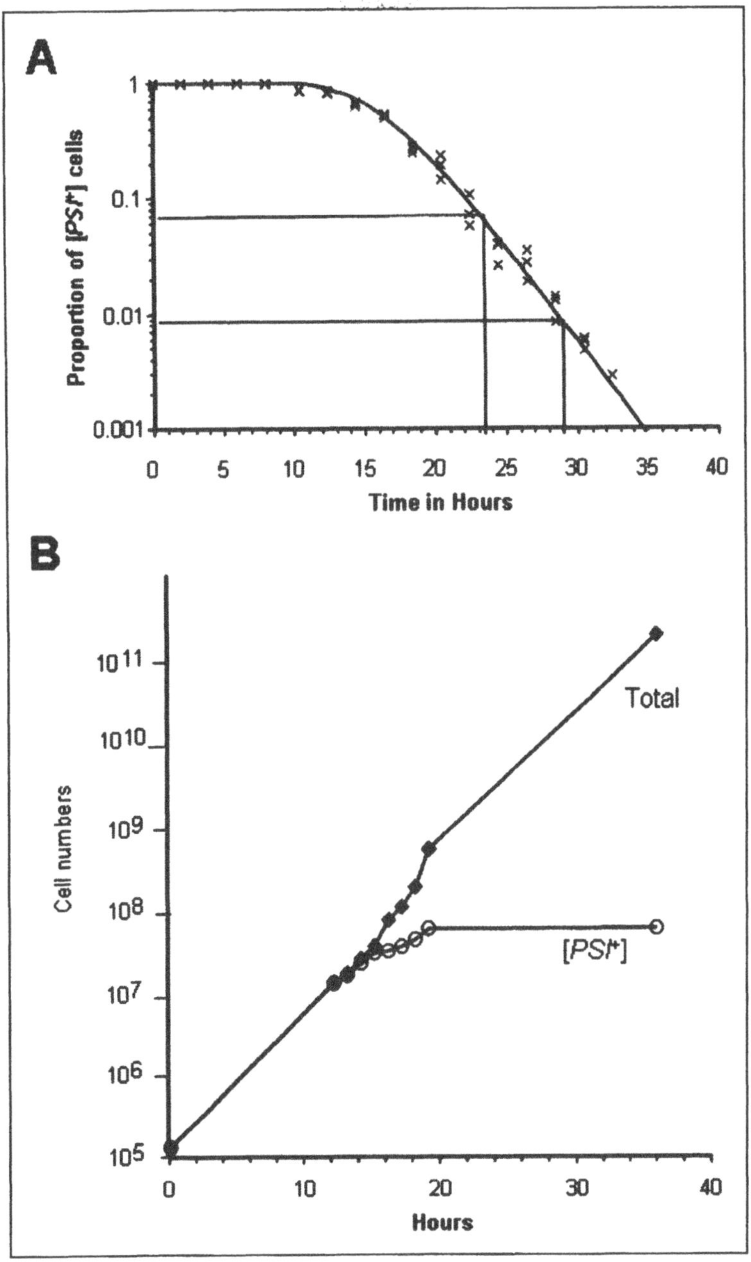

Figure 1. A) The proportion of ψ^+ cells left in a growing population following the addition of 5 mM guanidine hydrochloride. The slope of the exponential part of the fitted curve involves a halving of the proportion every 1.8 hours. This was derived from the doubling time of the cell numbers in this culture. The fitted line is from the model described by Cole et al.[43] B) Data from a curing experiment as described in 1A replotted to show that the numbers of ψ^+ cells in the culture reaches a maximum but then stays constant as the culture continues to grow at the normal rate (compare ref. 44).

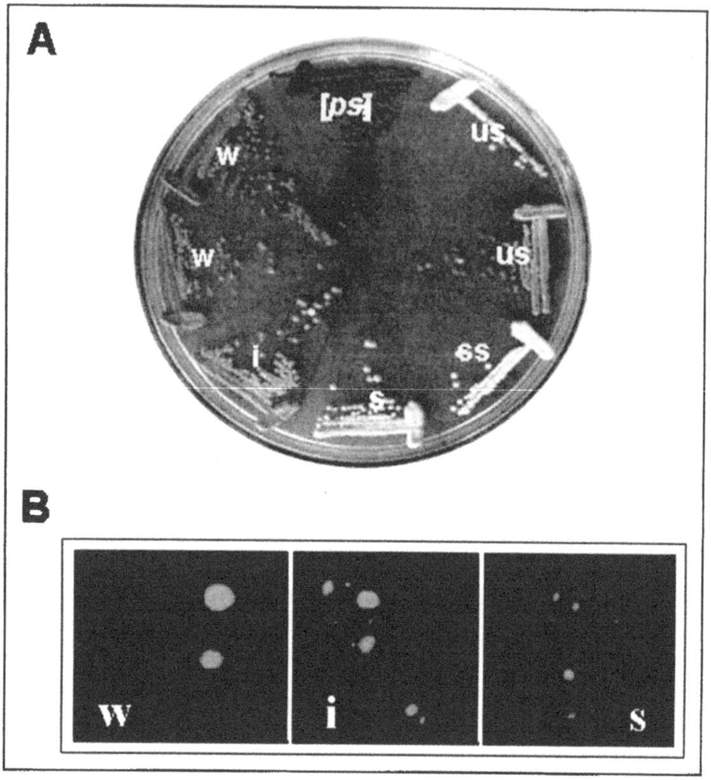

Figure 2. A) ψ^+ variants of strain 74-D694 identified by Eric Fernandez-Bellot. W, weak; i, intermediate; s, strong; ss super-strong; us, unstable. B) Sup35-GFP aggregates in different variants.

Models of Maintenance

Variants and the Steady State

Stability is a property of the particular ψ^+ "variant" being considered. So far, most people have observed a positive correlation between stability and the "strength" of a ψ^+ variant.[15] "Strength" is estimated by the level of suppression, usually by eyeball analysis of the development of the red colour that is due to deficiency in adenine biosynthesis. White is strong (Fig. 2). There has also been noted a positive correlation between the copy number of propagons and stability (Fig. 2 and Table 1). Thirdly, it has also been noted that weaker ψ^+ variants have larger SDS-resistant "oligomers" as defined by SDS-agarose electrophoresis[6,50] and, as expected, more soluble Sup35p.[57] It is clear there is an inter-relationship between these four properties: number and size of oligomers, read-through of stop codons and stability in mitosis.

Tanaka et al have developed an elegant statistical model which convincingly correlates them.[6] There are three quantities, namely the concentrations of monomeric (functional) Sup35p [x], the mass of aggregated Sup35p [z] and the concentration, that is numbers, of fibres or aggregates [y]. In real life, [x] determines the level of suppression: the more functional monomer there is, the less read-through of stop codons; and [y] is the copy-number, which affects stability. These values are governed by four rate constants, namely α, the rate of synthesis of Sup35p; (β) the rate of fibre growth, i.e., the rate at which monomers are

Table 1. Numbers of propagons in ψ^+ variants identified by Eric Fernandez-Bellot (see Fig. 2)

Strain	Propagons Count
Weak 2	70 ± 10
Weak1	105 ± 14
Intermediate	126 ± 15
Strong	171 ± 21

incorporated into aggregates (which includes and involves the conversion to the prion form); (γ), the rate at which aggregates are broken down to smaller sizes (by the action of Hsp104) and the rate of cell growth (R) (Fig. 3). β and γ are relevant only to the ψ^+ states and are, as determined by these workers' earlier experiments, determined by the structural features peculiar to each ψ^+ variant.[31]

Equations are derived which define steady-state conditions for the ψ state, when all the Sup35p is present as functional monomer, and for ψ^+ states which significantly reduce the amount of monomer by maintaining a significant proportion of the Sup35p as aggregate.

They find, interestingly, that the ψ^- steady state is not stable: it only persists because of the very high kinetic barrier to spontaneous folding or refolding to the prion form of the protein. (Natural or artificial prion proteins other than Sup35p might, in some cases, have much lower intrinsic stability in the nonprion form). This stability persists only as long as there is no prion

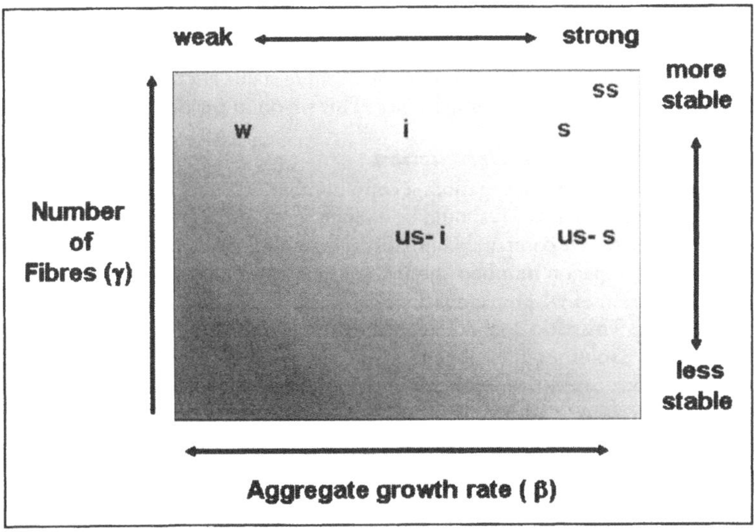

Figure 3. A diagram of the parameters determining ψ^+ variant strength and stability: adapted from Figure 1B in Tanaka et al, ref. 6. Strength is determined by the concentration of monomeric Sup35p: the less there is, the stronger the suppression. In ψ^+ strains that is determined by the growth rate of aggregate (β) relative to the rate of synthesis of Sup35p (α). Stability is related to the numbers of fibres (vertical axis) which is determined by their fragmentation by Hsp104 etc. (γ). Possible position of the variants found by Fernandez-Bellot (Fig. 2) are suggested. Tanaka et al suggest "real" positions of the three variants they describe, calculated from their measured parameters of fibre growth (β), fragility (γ) and number.

conformer present: once present, β (fibre growth) and γ, fibre division ensure a ψ^+ steady state. Tanaka determine that ψ^+ steady states are stable, but this is probably because they make the assumption in developing their equations (in the absence of contrary evidence) that aggregate degradation to nonprion monomers is negligible in vivo. The actual amounts of monomer, aggregate and numbers of aggregates (fibres) defining variant strength, stability and the size of aggregates, can vary with β and γ independently over quite wide ranges but still tolerate a considerable spectrum of relatively stable ψ^+ variants, and β and γ are properties determined for any variant by its structure. Stability is then solely a property of fibre number and partition.

The kinetic instability of the ψ^- state is a mathematical way of defining a prion as a molecular state requiring a template for its propagation. Tanaka et al ask whether ψ requires one, two, three or four molecules for a template and find from their infection assay that one will do (ref. 6, Supplementary information). Their analysis is supported by a comparison of three phenotypically different ψ^+ variants and they relate the measured properties of the variants (propagon number, level of suppression, in vitro fibre fragility and growth rate, and in vivo regeneration rates) to their phenotypes.

The beauty of this analysis is that it allows one to recognize the properties of prions in all their variety as manifestations of a few definable, measurable properties. For example, the strange variant described by Borchsenius et al[58] can be understood as a conformer of Sup35p which, its OR region being perhaps rather inaccessible to Hsp104, has a very low γ rate constant, and so occupies a position on the phanerogram in Figure 3, low down and perhaps towards the right-hand edge. The analysis also allows and accounts for the occurrence of variants which do not exhibit the conventionally accepted correlations of, for example, "strong" with "stable". In this figure, adapted from Figure 1B in Tanaka et al, we suggest with no experimental evidence apart from suppression and stability, positions for some of the variants described by Fernadez-Bellot (Fig. 2). In their paper, however, Tanaka et al define positions on the diagram for three variants based on measurements of both the β and γ rate constants.

While this elegant analysis defines the metabolic rate constants and concentration conditions needed to provide the foundations for establishing stable alternative prion states, it stops short of discussing mechanisms of inheritance. This we do in the next section.

Partition, Feed-Back and Destruction

Given genuinely random segregation, a copy-number of 20 such as that found for one of Tanaka et al's weak variants, Sc37, should be enough for ψ^- segregants to appear at a frequency of only 10^{-6} x generation^{-1}: comparable to most gene mutation rates. Other variants with typically much higher propagon numbers should be much more stable. However, Cox et al, measuring the distribution of ψ^+ propagons at cell division found that mothers acquired twice as many as daughters.[7] This ratio does not necessarily imply nonrandomness: it is directly proportional to the relative volumes of mother and daughter at cytokinesis. Nevertheless, if perpetuated, the discrepancy could lead to much greater instability than with 50:50 partition. Five generations of successive malsegregations of this degree without correction would generate about 3% of cells with one or fewer propagons, and it gets worse. Instability of this order is not observed, even in weak variants. Correction may be inherent in the physiology of yeast cell division. Lee Byrne has observed that daughter cells take twice as long to enter the next round of cell division as do their mothers and if propagon synthesis is continuous, that would neatly balance the discrepancy in numbers from unequal partition (L.J. Byrne, unpublished observations).

Conditions can arise, or be imposed, which affect partition. Ness et al report that when curing by guanidine HCl occurs, towards the end of the lag period partition becomes very inefficient and daughters seldom inherit ψ^+: the prion or prions stay in the mother in >90% of cell divisions[7,42] (Fig. 4). This is understandable in terms of the Tanaka model, since inactivation of Hsp104 affects only γ (fibre division) but not, as Ness et al show, the other two rate

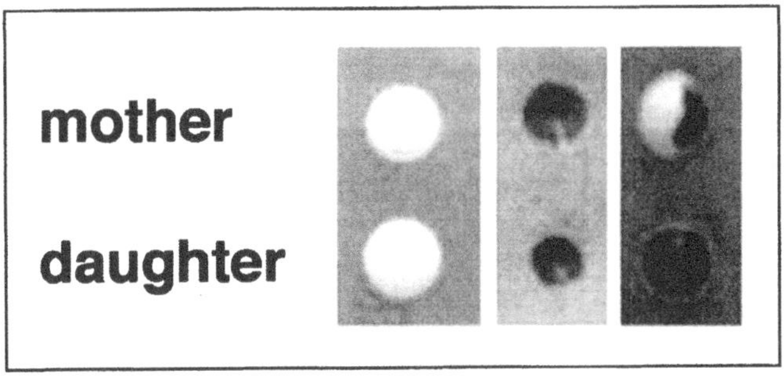

Figure 4. Segregation of ψ^+ and ψ^- in mother-daughter pairs taken from a culture growing in guanidine hydrochloride in which only 67% of the cells remained ψ^+. Budded cells were selected by micromanipulation on a YEPD agar plate and the daughter bud separated from the mother, each being left to grow into a colony at a marked place on the plate. The third pair illustrates segregation of ψ^+ from ψ^- both in the chosen pair of cells and at the next division of the mother cell. It is evident that the ψ^+ prion remained in the mother at the first division, as it did in 93% of the mother daughter pairs taken from this culture.

constants. Thus, fibres continue to grow under the curing regime, and after several generations may reach a size which interferes with their passage to daughter cells.

Similarly, partition may be affected by the over-expression of Hsp104 itself. The ψ prion is unique in its destabilization by high levels of Hsp104: no other yeast prion is affected by this treatment. It is a common assumption that excess Hsp104 breaks down ψ aggregates and so destroys propagons. However, it is also observed that serious partition defects appear in these conditions and these could well explain the curing effect (Fig. 5). Fernandez-Bellot also showed

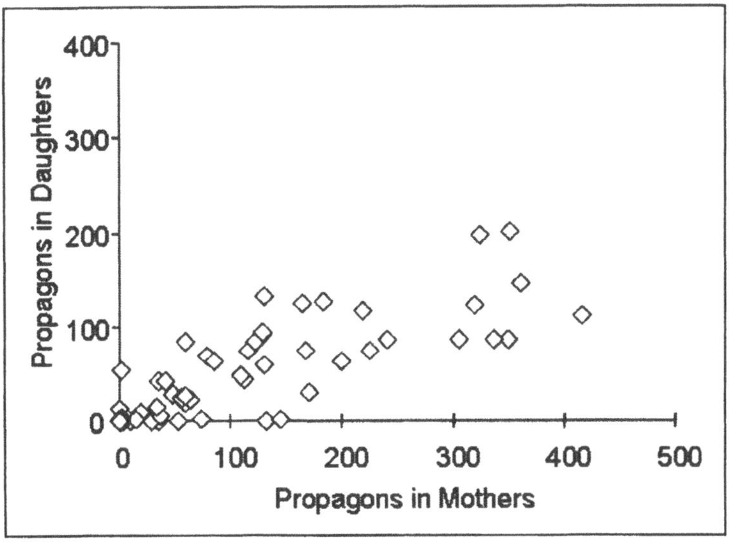

Figure 5. Propagons in pairs of mother and daughter cells taken from a culture of ψ^+ cells in which Hsp104 had been over-expressed for 20 hours. The mothers and daughters were separated after cytokinesis and placed on 3 mM guanidine hydrochloride as described in reference 6. The figure illustrates malsegregation in a minority of the dividing cells, never observed in normal cultures. 36% of the colonies from this culture were ψ^- or sectored.

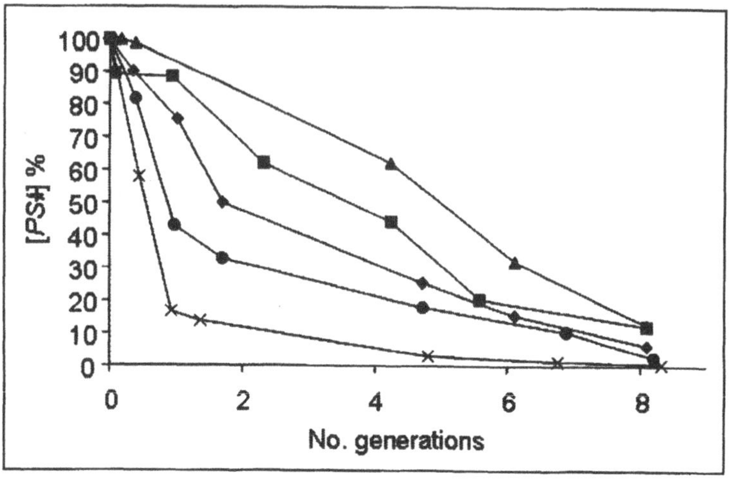

Figure 6. Weak variants are more readily de-stabilized by over-expressed Hsp104 than are strong variants. The symbols correspond to: ▲, strong; ■, unstable; ◆, intermediate; ● and X, weak.

that sensitivity of ψ^+ variants to excess Hsp104 was inversely proportional to their strength (Fig. 6) and this is perhaps more consistent with a partition defect, it being more apparent in low copy-number variants with larger aggregates (Fig. 6; Fernandez-Bellot, unpublished observations).

There is no indication, apart from these few departures from normal distribution in unusual circumstances, that there may be any mechanism that promotes equal partition of propagons of ψ^+.

Equally, evidence of feed-back systems maintaining numbers of propagons sufficient for stability is light. Tanaka et al state that stability of ψ^+ is achieved when $\alpha\beta\gamma/R^3$ is equal to or greater than 1.[6] However, ratios greater than 1 would imply an accumulation with time of both monomer and aggregate, whereas it is apparent that the totals stay quite constant. If the ultimate regulator is the rate of synthesis of monomer, then the only feed-back is from, and inherent in, those systems coupling protein synthesis with cell growth and division.

In experiments that measure the rate of synthesis of new propagons in cells depleted of them by growth in guanidine HCl, Ness et al have found that there is an exponential increase in numbers with a doubling time of about 20 minutes and that the numbers level out after about two hours.[42] This would appear to suggest a feed-back control of some sort, but Tanaka et al show that it is a consequence of the balance between the four rate constants as they define them and the saturation of the terms. The kinetics of the recovery of propagon numbers that they calculate from estimates of the rate constants for a strong ψ^+ variant closely match those observed by Ness et al (ref. 6, supplementary material).

The Effects of Chaperones: Prion Degradation

Although it alone is essential for the replication of prions, the chaperone Hsp104 interacts with both Hsp70 and Hsp40 chaperones in prion maintenance. Chapter 8 by Rikhvanov et al in this volume discusses their roles and link them in a schematic model involving them in generation, reproduction and inactivation of the ψ prion.[53] Recently conditions have been described which suggest that propagons may be lost by means other than inactivation of their reproduction and dilution by cell division.[54] These authors found that during a prolonged inhibition of cell growth by α-factor in the presence of guanidine HCl, curing of ψ^+ with kinetics apparently similar to those observed when cured during growth could be observed.

They suggest that this is brought about by destabilisation of prion aggregates as a result of their growth unfettered by the activity of Hsp104.

Although Wu et al record no observations of such destabilisation, it is entirely possible that in these highly abnormal conditions prion reproductive activity may be lost by their degradation, sequestration or occlusion, and Rikhvanov et al (loc. cit.) include such a pathway in their model, exploiting the known properties of chaperones in disaggregating, refolding or degrading nonnative proteins. For [URE3$^+$] also, two systems by which it might be cured to the [ure3-0] state have been described.[47] One, as with [ψ^+], is brought about by growth in guanidine hydrochloride and is accounted for by dilution through cell division. The other is observed when the Ure2p PFD (residues 1-93) is over-expressed. If this is done in [URE3$^+$] cells, large aggregates are formed (containing both full-length protein and PFD) concomitantly with the loss of [URE3$^+$] cells. The aggregates evidently have no propagon activity: they are to all intents and purposes dead and whether they are passed to daughter cells or not, they are not destroyed over a period of 40 hours. Quite possibly this also happens to some extent with Sup35, since blocking Hsp104 activity still allows soluble Sup35p to accrete to aggregates and we have noted above the effect of these treatments on partition (ref. 42 and Fig. 4). This effect may be exaggerated in α-factor arrested cells where dispersion of large aggregates cannot be ameliorated by cell division. It has yet to be established what is going on in the yeast strain treated as is described by Wu et al.[54] Whatever it is it clearly plays a minor role in normally dividing cells.

Summary

The Sup35 protein of yeast has several hyperstable states: ψ^-, in which all the Sup35 protein in the cell is in a soluble native form, active as the only and therefore essential, eRF3 in translation; and prion variants of ψ^+ in which up to 95% of the protein is in an aggregated form unable, as far as is known, to function in polypeptide chain termination. There are, in addition, conditions in which nonstable states may exist, which are not heritable. For example, when Sup35 or C-truncated versions of it are over-expressed, high molecular weight oligomeric aggregates can be detected in cell cultures and it is speculated that there may be soluble monomeric or oligomeric intermediates formed transiently in the conversion of the native monomer to ψ^+ forms. Conversion in vivo of ψ^- into any of the ψ^+ variants requires the presence of another heterologous prion, commonly that of the Rnq1 protein; or of amyloid. This is also true of the nonheritable aggregates produced by over-expression.

The stable ψ^- conformer may owe its stability to the intrinsically high entropy characteristic of folded functionally evolved proteins, but this is overcome, with some difficulty by interactions, whose nature is unknown, with heterologous prions; and with great ease when any variant ψ^+ conformer is present. A single prion molecule is sufficient to initiate the switch from the stable ψ^- state of a cell, which contains 100% of soluble active monomers, to a stably ψ^+ cell with 95% of those molecules being in prion form, in aggregates large or small.

This is the basis of the chemical stability of ψ^- and ψ^+ conditions. ψ^+ stability in dividing cells needs further conditions, since the maintenance of the state in vivo critically depends on

[e] It should of course be noticed that this condition, by its absence, is what maintains the stability of the ψ^- state and defines prions. If there were no requirement for a template, there would be no "switch", only a distribution, potentially in every cell, between forms depending on the energetic parameters of each form, the proportions being conditioned by their relative entropies. These in turn may depend on sequence. For example, the propensity of huntingtin to form amyloid is directly related to the number of glutamine residues in the polyQ region. There might also be intermediate situations, in which the likelihood of amyloid formation depends on the previous presence of amyloid: this would be prion-like, at least as far as the conversion part of the process was concerned. Inheritance or infection, however, would depend in addition on reproduction and transmission of the amyloid. It might be quite difficult in some cases to distinguish between the necessity of a template, defining a prion and mere kinetic instability. [*PHI*$^+$], might for example be one such case.[30]

the presence of a pre-existing prion molecule.[e] Firstly, stability requires that such "seeds" or propagons interact with nonprion native conformers to promote their conversion. Secondly, it requires that propagons are inherited by both progeny of every cell division and this means that their numbers have to be maintained at the same rate at which the cells divide and that they be partitioned efficiently. At least two of these conditions depend on apparently distinct sequences in the prion-forming domain of the proteins. Interaction of propagons and native monomer, identified in biochemical assays as the formation of aggregates, depends in Sup35 on the QN-rich region in the first 40 residues at the N-terminal. This is indicated by the fact that this is the domain specifying the "species barrier" and also by the fact that it can be substituted without much affecting ψ^+ propagation by known aggregation-promoting sequences from heterologous sources or by synthetic sequences. The maintenance of propagon numbers depends on the activity of Hsp104 interacting with another domain in the Sup35 protein, namely the oligopeptide repeat region. Again there are heterologous sequences from other prion-forming proteins that can substitute for this function.

Acknowledgement

B.S.C. wishes to acknowledge the award of an Emeritus Fellowship by the Leverhulme Trust, during the tenure of which this work was done.

References

1. Coustou V, Deleu C, Saupe S et al. The protein product of the het-s heterokaryon incompatibility gene of the fungus Podospora anserina behaves as a prion analog. Proc Natl Acad Sci USA 1997; 94:9773-9778.
2. Wickner RB. [URE3] as an altered URE2 protein: Evidence for a prion analog in Saccharomyces cerevisiae. Science 1994; 264:566-569.
3. Cox BS. Prion-like factors in yeast. Curr Biol 1994; 4:744-748.
4. Derkach IL, Bradley ME, Hong JY et al. Prions affect the appearance of other prions: The story of [PIN⁺]. Cell 2001; 106:171-182.
5. Osherovich LZ, Weissman JS. Multiple Gln/Asn-rich prion domains confer susceptibility to induction of the yeast [PSI⁺] prion. Cell 2001; 106:183-194.
6. Tanaka M, Collins SR, Toyama BH et al. The physical basis of how prion conformations determine strain phenotypes. Nature 2006; 442:585-589.
7. Cox BS, Ness F, Tuite MF. Analysis of the generation and segregation of propagons: Entities that propagate the [PSI⁺] prion in yeast. Genetics 2003; 165:23-33.
8. Lund PM, Cox BS. Reversion analysis of [psi⁻] mutants in Saccharomyces cerevisiae. Genet Res 1981; 37:173-182.
9. Juliani MH, Gambarini AG, Da Costa MOP. Induction of rho-minus mutants in Saccharomyces cerevisiae by guanidine hydrochloride. I. enetic analysis. Mutation Res 1975; 29:67-75.
10. Derkach IL, Bradley ME, Zhou P et al. Genetic and environmental factors affecting the de novo appearance of the [PSI⁺] prion in Saccharomyces cerevisiae. Genetics 1997; 147:507-519.
11. Masison DC, Maddelein ML, Wickner RB. The prion model for [URE3] of yeast: Spontaneous generation and requirements for propagation. Proc Natl Acad Sci USA 1997; 94:12503-12508.
12. Bradley ME, Edskes HK, Hong JY et al. Interactions among prions and prion "strains" in yeast. Proc Natl Acad Sci USA 2002; 99(suppl. 4):16392-16399.
13. Chernoff YO, Derkach IL, Inge-Vechtomov SG. Multicopy SUP35 gene induces the de novo appearance of psi-like factors in the yeast Saccharomyces cerevisiae. Curr Genet 1993; 24:268-270.
14. Sondheimer N, Lindquist S. Rnq1: An epigenetic modifier of protein function in yeast. Molec Cell 2000; 5:163-172.
15. Derkach IL, Chernoff YO, Kushnirov VV et al. Genesis and variability of [PSI] prion factors in Saccharomyces cerevisiae. Genetics 1996; 144:1375-1386.
16. Masison DC, Wickner RB. Prion-inducing domain of yeast Ure2p and protease resistance of Ure2⁻ in prion-containing cells. Science 1995; 270:93-95.
17. Roberts BT, Wickner RB. A new kind of prion. Cell Cycle 2004; 3:100-103.
18. Derkach IL, Bradley ME, Masse SVL et al. Dependence and independence of [PSI⁺] and [PIN⁺]: A two-prion system in yeast? EMBO J 2000; 19:1942-1952.
19. Derkach IL, Uptain SM, Outeiro TF et al. Effects of Q/N-rich, polyQ and non-polyQ amyloids on the de novo formation of the [PSI⁺] prion in yeast and aggregation of Sup35 in vitro. Proc Natl Acad Sci USA 2004; 101:12934-12939.

20. Salnikova AB, Kryndushkin DS, Smirnov VN et al. Nonsense suppression in yeast cells overproducing Sup35 (eRF3) is caused by its nonheritable amyloids. J Biol Chem 2005; 280:8808-8812.
21. Kochneva-Pervukhova NV, Poznyakovski AI, Smirnov VN et al. C-terminal truncation of the Sup35 protein increases the frequency of de novo generation of a prion-based [PSI⁺] determinant in Saccharomyces cerevisiae. Curr Genet 1998; 34:146-151.
22. Fernandez-Bellot E, Guillemet E, Cullin C. The yeast prion [URE3] can be greatly induced by a functional mutated URE2 allele. EMBO J 2000; 19:3215-3222.
23. Hatin I, Bidou L, Cullin C et al. Translational errors as an early event in prion conversion. Cell Mol Biol (Noisy-le-grand) 2001; 47:23-28, (Online Pub. OL).
24. Kolotova-Levine N, Merritt G, Tuite MF. Unpublished results.
25. Liu JJ, Lindquist S. Oligopeptide-repeat expansions modulate 'protein-only' inheritance in yeast. Nature 1999; 400:573 -576.
26. Parham SN, Resende CG, Tuite MF. Oligopeptide repeats in the yeast protein Sup35p stabilize intermolecular prion interactions. EMBO J 2001; 20:2111-2119.
27. De Pace AH, Santoso A, Hillner P et al. A critical role for amino-terminal glutamine/asparagine repeats in the formation and propagation of a yeast prion. Nature Struct Biol 2002; 9:389-396.
28. Santoso A, Chien P, Osherovich LZ et al. Molecular basis of a yeast prion species barrier. Cell 2000; 100:277-288.
29. Osherovich LZ, Cox BS, Tuite MF et al. Dissection and design of yeast prions. PLoS Biology 2004; 2:0442.
30. Crist CG, Nakayashiki T, Kurahashi H et al. [PHI⁺], a novel Sup35-prion variant propagated with non-Glu/Asn oligopeptide repeats in the absence of the chaperone protein, Hsp104. Genes and Cells 2003; 8:603-618.
31. Tanaka M, Chien P, Naber N et al. Conformational variations in an infectious protein determine prion strain differences. Nature 2004; 428:265-267.
32. King CY, Diaz-Avalos R. Protein-only transmission of three yeast prion strains. Nature 2004; 428:319-323.
33. Satpute-Krishnan P, Serio TR. Prion protein remodelling confers an immediate phenotypic switch. Nature 2005; 437:262-265.
34. Mehta S, Yang XM, Chan CS et al. The 2-micron plasmid purloins the yeast cohesin complex: A mechanism for coupling plasmid partitioning and chromosome segregation. J Cell Biol 2002; 158:625-637.
35. Chernoff YO, Lindquist SL, Ono B et al. Role of the chaperone protein Hsp104 in propagation of the yeast prion-like factor [psi⁺]. Science 1995; 268:880-884.
36. Tuite MF. D. Phil Thesis. Oxford: 1979.
37. Tuite MF, Mundy CJ, Cox BS. Agents that cause a high frequency of genetic change from [psi⁺] to [psi⁻] in Saccharomyces cerevisiae. Genetics 1981; 98:691-711.
38. Glover JR, Lindquist SL. Hsp104, Hsp70 and Hsp40: A novel chaperone system that rescues previously aggregated proteins. Cell 1998; 94:73-82.
39. Ferreira PC, Ness F, Edwards SR et al. The elimination of the [psi⁺] prion by guanidine hydrochloride is the result of Hsp104 inactivation. Mol Microbiol 2001; 40:1357-1369.
40. Jung G, Masison DC. Guanidine hydrochloride inhibits Hsp104 activity in vivo: A possible explanation for its effect in curing yeast prions. Curr Microbiol 2001; 43:7-10.
41. Eaglestone S, Ruddock LW, Cox BS et al. Guanidine hydrochloride blocks a critical step in the propagation of the prion-like determinant [PSI⁺] of Saccharomyces cerevisiae. Proc Natl Acad Sci USA 2000; 97:240-244.
42. Ness F, Ferreira P, Cox BS et al. Guanidine hydrochloride inhibits the generation of prion seeds but not prion protein aggregation in yeast. Mol Cell Biol 2002; 22:5593-5605.
43. Cole DJ, Morgan BJT, Ridout MS et al. Estimating the number of prions in yeast cells. Mathematical Medicine and Biology 2004; 21:369-395.
44. Cox BS. In: Hall MD, Linder P, eds. Psi Phenomena in Yeast: Early Days of Yeast Genetics. NY: Cold Spring Harbor Laboratory Press, 1993:477.
45. Kushnirov VV, Ter-Avenasyan MD. Structure and replication of yeast prions. Cell 1998; 94:13-16.
46. Wegrzyn RD, Bapat K, Newnam GP et al. Mechanisms of prion loss after Hsp104 inactivation in yeast. Molec Cell Biol 2001; 21:4656-4669.
47. Ripaud L, Maillet L, Cullin C. The mechanisms of [URE3] prion elimination demonstrate that large aggregates of Ure2p are dead-end products. EMBO J 2003; 22:5251-5259.
48. Singh A, Helms C, Sherman F. Mutation of the non-Mendelian suppressor, Psi, in yeast by hypertonic media. Proc Natl Acad Sci USA 1979; 76:1952-1956.
49. Tuite MF, Cox BS. Ultraviolet mutagenesis studies of [psi], a cytoplasmic determinant of Saccharomyces cerevisae. Genetics 1980; 95:611-630.

50. Kryndushkin DS, Alexandrov IM, Ter-Avanesyan MD. Yeast [PSI+] prion aggregates are formed by small Sup35 polymers fragmented by Hsp104. J Biol Chem 2003; 278:49636-49643.
51. Liebman SL, Derkach IL. Prion-Prion interactions. 2007; this volume.
52. Tuite MF, Cox BS. The genetic control an propigation of the [PSI+] prion in yeast. 2007; this volume.
53. Rikhvanov EG, Romanova NV, Chernoff YO. Chaperone effects on prion and nonprion aggregates. 2007; this volume.
54. Wu YX, Greene LE, Masison DC et al. Curing of yeast [PSI+] prion by guanidine inactivation of Hsp104 does not require cell division. Proc Natl Acad Sci USA 2005; 102:12789-12794.
55. Borchsenius AS, Wegrzyn RD, Newnam GP et al. Yeast prion protein derivative defective in aggregate shearing and production of new 'seeds'. EMBO J 2001; 20:6683-6691.
56. Futcher AB. The 2 micron circle plasmid of Saccharomyces cerevisiae. Yeast 1988; 4:27-40.
57. Zhou P, Derkach IL, Uptain SM et al. The yeast non-Mendelian factor [ETA+] is a variant of [PSI+], a prion-like form of release factor eRF3. EMBO J 1999; 18:1182-1191.
58. Borchsenius AS, Muller S, Newnam GP et al. Prion variant maintained only at high levels of the Hsp104 disaggregase. Curr Genet 2006; 49:21-29.

Chapter 6

Prion and Nonprion Amyloids:
A Comparison Inspired by the Yeast Sup35 Protein

Vitaly V. Kushnirov, Aleksandra B. Vishnevskaya, Ilya M. Alexandrov and Michael D. Ter-Avanesyan*

Abstract

Yeast prion determinants are related to polymerization of some proteins into amyloid-like fibers. The [*PSI*$^+$] determinant reflects polymerization of the Sup35 protein. Fragmentation of prion polymers by the Hsp104 chaperone represents a key step of the prion replication cycle. The frequency of fragmentation varies depending on the structure of the prion polymers and defines variation in the prion phenotypes, e.g., the suppressor strength of [*PSI*$^+$] and stability of its inheritance. Besides [*PSI*$^+$], overproduction of Sup35 can produce nonheritable phenotypically silent Sup35 amyloid-like polymers. These polymers are fragmented poorly and are present due to efficient seeding with the Rnq1 prion polymers, which occurs by several orders of magnitude more frequently than seeding of [*PSI*$^+$] appearance. Such Sup35 polymers resemble human nonprion amyloids by their nonheritability, mode of appearance and increased size. Thus, a single protein, Sup35, can model both prion and nonprion amyloids. In yeast, these phenomena are distinguished by the frequency of polymer fragmentation. We argue that in mammals the fragmentation frequency also represents a key factor defining differing properties of prion and nonprion amyloids, including infectivity. By analogy with the Rnq1 seeding of nonheritable Sup35 polymers, the "species barrier" in prion transmission may be due to seeding by heterologous prion of nontransmissible type of amyloid, rather than due to the lack of seeding.

Prion and Nonprion Amyloids of Mammals

Prions were originally defined as a unique class of infectious agents, whose infectivity relates solely to protein. In mammals, they cause fatal neurodegenerative diseases, such as Creutzfeldt-Jacob disease of man, sheep scrapie and bovine spongiform encephalopathy (for review, see refs. 1,2). All these diseases are related to the PrP protein, whose conformationally altered form (PrP^{Sc}) is able to convert the normal host-encoded protein (PrP^{C}) into this altered prion form. While only one prion protein is known in mammals, the prions appear to represent just a part of a much wider phenomenon, amyloidoses.

Amyloid diseases represent a group of more than 20 human diseases, which are characterized by deposition in different tissues of fibrous aggregates of conformationally altered proteins.[3] Some of these diseases, like Alzheimer's and Parkinson's disease, represent a major challenge for the public health care in the developed countries. Although amyloidogenic proteins are structurally and functionally unrelated, they form morphologically similar amyloid fibers which reproduce the key properties of prions: the normal and polymer forms of amyloidogenic

*Corresponding Author: Michael D. Ter-Avanesyan—Institute of Experimental Cardiology, Cardiology Research Center, 3rd Cherepkovskaya Street 15A, Moscow 121552, Russia. Email: mdter@cardio.ru

Protein-Based Inheritance, edited by Yury O. Chernoff. ©2007 Landes Bioscience.

proteins are structurally different and the latter can promote polymerization of normal proteins into amyloid fibers. Unlike prion diseases, amyloid diseases are generally not transmissible.

Lately, the prion and amyloid phenomena acquired a more general significance for biology due to the finding of prions in lower eukaryotes and observations that in multicellular organisms they may be used in important biological mechanisms. It was shown that amyloid polymers of Pmel17 template and accelerate the polymerization of melanin precursor into melanin.[4] Melanin precursor is toxic, while its polymers protect cells from a broad range of cytotoxic insults including UV and oxidative damage. A very important candidate for prion-like mechanism relates to long-term memory in animals. The translational regulator CPEB of *Aplysia californica*, which plays a key role in the long-term synaptic changes associated with memory storage, demonstrated prion-like properties in yeast.[5]

Yeast Prions

Prion-like protein behavior underlies the inheritance of some phenotypic traits in the yeast *Saccharomyces cerevisiae*[6] and filamentous fungus *Podospora anserina*.[7] In yeast, there are several proteins, which can undergo prion-like structural conversion. The most studied of them are translation termination factor eRF3, also called Sup35, and Ure2 involved in regulation of nitrogen metabolism.[6] The prion state of these proteins can propagate stably for many cellular generations. This may be observed by characteristic phenotypes, [*PSI*$^+$] and [*URE3*], which reflect reduced function of the respective proteins, Sup35 and Ure2, due to aggregation of their prion form. In particular, the prion state of Sup35 results in low levels of soluble functional Sup35 and impaired translation termination, which is manifested as a nonsense-suppressor [*PSI*$^+$] phenotype. Unlike [*PSI*$^+$] and [*URE3*], the phenotypic manifestation of the third yeast prion, [*PIN*$^+$], is not related to inactivation of the corresponding protein. This prion determinant was originally described as a factor required for the [*PSI*$^+$] de novo generation by transient Sup35 overproduction.[8] Later it was found that it also facilitates the de novo appearance of [*URE3*].[9] Unlike the appearance, propagation of [*PSI*$^+$] and [*URE3*] does not depend on the presence of [*PIN*$^+$].

The prion properties of the described yeast proteins rely on their areas rich in glutamine and/or asparagine (QN), which apparently reflects an increased ability of QN-rich sequences to form prions. Analysis of selected yeast proteins with QN-rich regions uncovered Rnq1 prion, but other candidates showed only some of prion properties. When overproduced, the Ybr016w and Hrp1 proteins form aggregates evident by coalescence of their GFP fusions, but their prion abilities were not reported.[10] QN-rich region of the New1 protein formed prion when fused to Sup35, but prion formation by New1 itself was not shown.[11] A screen for a protein responsible for the [*PIN*$^+$] phenotype revealed Rnq1, Ure2, New1 and nine other candidate proteins. Overproduction of these proteins, together with Sup35, allowed [*PSI*$^+$] appearance in [*pin*$^-$] [*psi*$^-$] cells.[12] This relied on the observation that overproduction of a prionogenic protein greatly increases the probability of its switch into prion form. In this experiment, a prion-like switch of a candidate protein facilitated an analogous switch of Sup35. All candidates possessed QN-rich regions, but only Rnq1 and Ure2 are known to exist in a prion mode. Possibly, other candidates aggregate stably only when they are overproduced. [*PIN*$^+$] in most cases is related to the prion state of Rnq1, for which Rnq1 is usually implied as a protein underlying the [*PIN*$^+$] determinant. However, it should be kept in mind that the Ure2 prion also exhibits the Pin$^+$ phenotype. Interestingly, both [URE3] and [*PSI*$^+$] facilitate the appearance of the prion form of Rnq1,[12] when Rnq1 is overproduced. Thus, QN-rich prions promote one another's de novo appearance.

Modular Structure of Yeast Prion Proteins

The Sup35 protein has a clear three-domain structure. The nonessential N-terminal domain is responsible for [*PSI*$^+$] appearance and maintenance.[13,14] The charged middle (M) domain, is important, though not required for [*PSI*$^+$] propagation.[15,16] The C-terminal

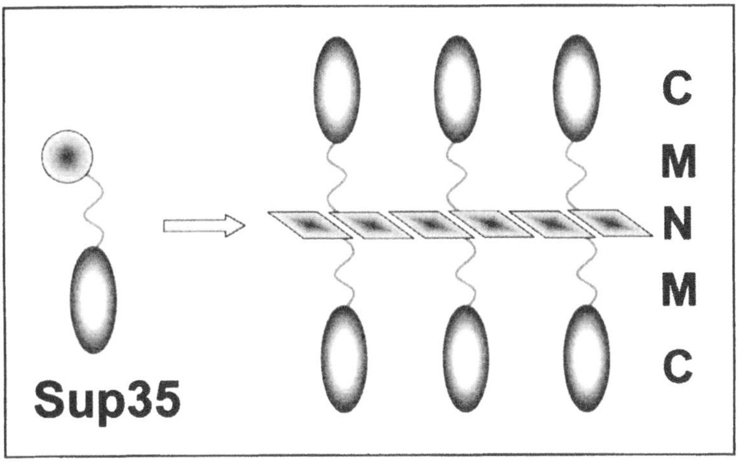

Figure 1. Structure of Sup35 polymers. Amyloid-like fiber is formed by the Sup35N domains. N, M and C, domains of Sup35.

domain performs the essential translation termination activity.[17] In [*PSI*⁺] cells, Sup35 polymerizes via its prion domain, which strongly reduces its function in termination (Fig. 1). The levels of soluble Sup35 decrease, in some cases to less than 1% of the [*psi*⁻] level.[18,19] This causes increased nonsense codon readthrough, which may be conveniently detected using the *ade2-1* UAA or *ade1-14* UGA nonsense mutations. In these mutants the [*psi*⁻] cells are adenine requiring and accumulate red pigment related to impaired adenine biosynthesis. [*PSI*⁺] cells are adenine-independent and form white (strong suppressor [*PSI*⁺] variants) or pink (weak variants) colonies.

The Ure2 protein has a similar structure, which includes the N-terminal prion-forming domain and the C-terminal functional domain structurally similar to glutathione transferases.[20] In contrast to Sup35 and Ure2, Rnq1 lacks conservative or presumed functional domains. The prion-forming region involves C-terminal two thirds of Rnq1.[10]

The modular structure of Sup35 allows convenient adaptation of the yeast [*PSI*⁺] system to test the prionogenic potential of various proteins or protein domains. This is performed by replacing the Sup35 prion domain, via DNA manipulations, for polypeptide sequences of interest, and studying the ability of fusion proteins to cause the [*PSI*⁺]-like phenotype. In another type of fusion, the C domain of Sup35 is replaced for green fluorescent protein (GFP), which allows visualization of prion aggregates by the appearance of bright fluorescent foci.[21]

Two-Level Structure of Prion Aggregates

The available data strongly suggest that yeast prions are composed of amyloid-like fibers. In vitro, purified Sup35, Ure2 and Rnq1 form fibers of uniform structure, which share the key properties of amyloid fibers.[22-24] Furthermore, such fibers of Sup35 and Ure2 can transform yeast cells from [*psi*⁻] to [*PSI*⁺] and from [ure3-0] to [URE3] states, respectively.[25-27] A simple treatment was discovered, which allows distinguishing prion particles from the vast majority of other cellular protein complexes. Unlike these complexes, prion polymers of Sup35 and Rnq1 are insoluble in the presence of SDS at room temperature,[28,29] which is likely to represent a general property of amyloids.

The treatment of Sup35 aggregates with SDS reduced their size about 35-fold. This led to the conclusion that the aggregates contain multiple SDS-resistant particles, prion polymers, which are likely to represent amyloid-like fibers. In addition, the aggregates should contain a significant amount of associated nonprion proteins and complexes, since even in a [*psi*⁻] state a

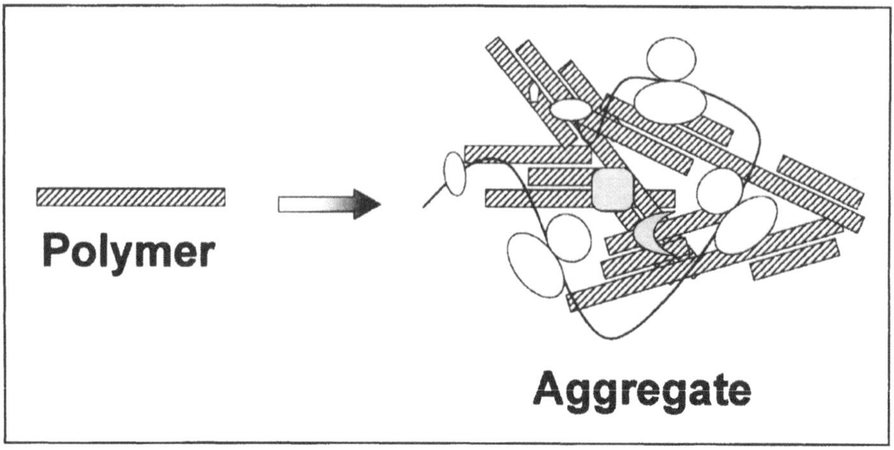

Figure 2. Prion polymers and aggregates. The aggregates represent irregular complexes containing multiple prion polymers and some additional proteins. In case of Sup35 these are presumably Sup35 functional partners, polyribosomes and chaperones.

large portion of Sup35 is found in high molecular mass fraction, being associated with polyribosomes[28] (Fig. 2).

A procedure for analysis of the size of prion polymers using SDS-agarose gels was developed, called semi-denaturing detergent-agarose gel electrophoresis (SDD-AGE). This method revealed that the Sup35 polymers of [PSI^+] cells comprise on average from 10 to 50 Sup35 molecules, depending on the prion variant.[28] Thus, yeast prion aggregates represent higher order complexes of relatively short amyloid-like polymers and some nonprion molecules.

Role of Hsp104 Chaperone in Yeast Prion Propagation

[PSI^+] propagation may be affected by chaperones of the Hsp70 and Hsp40 families (for review, see ref. 30), but the most significant role is played by Hsp104, which is shown to be strictly required for maintenance of [PSI^+][31] and other yeast prions.[8,32] Hsp104 breaks large aggregates of denatured protein into smaller pieces, which allows their further refolding by Hsp70 and Hsp40 and solubilization.[33,34] A similar action of Hsp104 applied to filamentous prion particles should yield a strikingly different effect. Every break of a filament should accelerate the prion conversion by creating new ends of prion polymers, where the conversion occurs, and multiply prion particles, which is required for their stable inheritance.[35] In fact, fragmentation completes the replication cycle of a prion. The other part of this cycle represents prion growth by polymerization, which does not require, at least in vitro, the help of additional factors (Fig. 3).

An alternative model for the role of Hsp104 was also proposed, presuming that Hsp104 facilitates prion conversion by helping to obtain some unfolded intermediate form of prionogenic protein.[21,36] However, studies of Hsp104 inhibition indicate strongly in favor of the former model. The activity of Hsp104 may be inhibited by growing yeast cells in the presence of low concentrations of guanidine-HCl (GuHCl).[37] This treatment is known to efficiently cure [PSI^+][38] and other known yeast prions. Study of the kinetics of [PSI^+] loss in the presence of GuHCl led to the conclusion that it blocks the replication of prion "seeds", named propagons,[39] while not interfering with incorporation of monomers into prions.[40]

Similar data were obtained using SDD-AGE analysis of the Sup35 polymer size.[28] Addition of GuHCl to medium caused a rapid increase in the size of polymers, while not interfering, at least for the first cell generation, with Sup35 polymerization. The size of Sup35 polymers grew twofold per cell generation, a rate that may only be achieved on conditions of the full block of

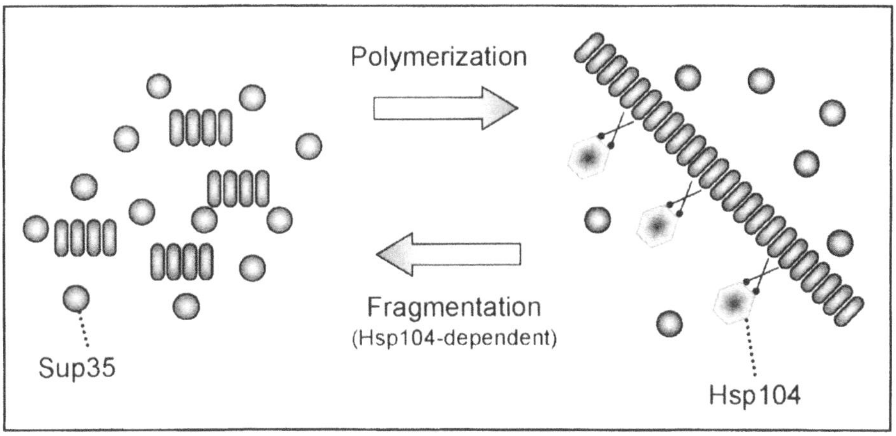

Figure 3. Replication of yeast Sup35 prion polymers. Polymers grow by joining Sup35 monomers and multiply by fragmentation with the Hsp104 chaperone.

fragmentation and complete incorporation of newly synthesized Sup35 into polymers. The [*PSI*⁺] curing by GuHCl could be prevented by mutations in the *HSP104* gene.[41] Most probably, this indicated that Hsp104 was solely responsible for the block, though did not rule out involvement of other proteins, as such a mutation could affect interaction of Hsp104 with other proteins. Repression of Hsp104 synthesis also increased the size of Sup35 polymers, although repression does not cause immediate inactivation of Hsp104. Thus, Hsp104 is required for polymer fragmentation, but does not affect polymerization.

Different Accessibility of Sup35 Polymers to Fragmentation Defines [*PSI*⁺] Prion Variability

Different variants or "strains" were observed for both mammalian and yeast prions. The variation was observed for [*PSI*⁺],[14] hybrid prion [PSI^{+}_{PS}] based on Sup35 prion domain from yeast *Pichia methanolica*,[42] [*URE3*][43] and [*PIN*⁺].[9] The [*PSI*⁺] variants are distinguished by the strength of nonsense suppressor phenotype and mitotic stability. Usually, stronger suppression correlates with higher stability. It was proposed[14,44,45] and recently confirmed[46,47] that the variation in [*PSI*⁺] properties reflects difference in the structure of prion particles. This may result in variation of prion polymerization speed and the frequency of fragmentation of prion polymers, and, therefore, in their different size. A comparison of the size of Sup35 polymers in different [*PSI*⁺] isolates indeed revealed a significant variation, with the size generally being inversely related to the strength of [*PSI*⁺]. The cells harboring strong [*PSI*⁺] have smaller polymers, which means that their number should be higher.[28] This explains both the higher mitotic stability of such [*PSI*⁺] and their stronger suppressor phenotype, which results from more efficient polymerization and lower levels of soluble functional Sup35.[18,19] However, this logic disregards possible variation of polymerization speed. If the speed varies significantly, the correlation of the polymer size and [*PSI*⁺] strength would be violated. Of the eight [*PSI*⁺] and [PSI^{+}_{PS}] variants tested, one variant violated the correlation: strong [$PSI^{+}_{PS\text{-}1}$] possessed large polymers.[28] This shows that the polymerization speed can vary significantly, but in most cases this variation may be neglected compared to variation in fragmentation frequency, which plays a dominant role in defining the variability of [*PSI*⁺] properties.

A putative uncertainty in correlation of the polymer number and stability of their inheritance should be noted. Prion stability depends on the number of prion seeds, or propagons, which probably correspond to prion aggregates, rather than to polymers. These considerations are supported by the effects of the *SSA1-21* mutation. This mutation, which alters the

Ssa1 chaperone, increases the size of Sup35 aggregates, and, therefore, decreases their number, while not affecting the Sup35 polymers.[48] It also greatly reduces [*PSI*⁺] stability, which shows that aggregates, rather than polymers, define prion inheritance. However, normally the numbers of polymers and aggregates correlate. Larger polymers seem to have higher propensity to aggregate, so an increase in the polymer size should cause an even greater increase in the size of prion aggregates.[28]

Nonprion Amyloids of Sup35

It may be proposed that some Sup35 amyloid structures possible in vitro would not propagate in vivo, being harmful to cells or nonheritable. Variants of [PSI^+] that are too strong would interfere with the cell viability due to the lack of soluble Sup35. At the other extreme, polymers may be fragmented too rarely, which would interfere with their heritability. Such unconventional Sup35 polymer variants may be uncovered in cells with altered Sup35 levels and/or structure.

Overproduction of Sup35 in cells with conventional [PSI^+] causes severe growth inhibition.[49] This may be related to the impairment of translation termination due to both the reduced levels of soluble Sup35 and titration of the Sup45 (eRF1) partner termination factor by functionally inactive Sup35 polymers. Any [PSI^+] compatible with Sup35 overproduction would represent a new prion variant not existing under standard conditions.

Two such novel types of [PSI^+] were obtained in a [PIN^+] strain overproducing Sup35: (i) [PSI^+] compatible with Sup35 overproduction, but stable at standard Sup35 levels; (ii) [PSI^+] existing only at increased Sup35 levels. In addition, a category of Sup35 amyloid-like polymers was discovered, which does not manifest itself as [PSI^+].[50] Surprisingly, these polymers were present in cells before the selection for suppressor phenotype, as revealed by SDD-AGE. Cells containing these polymers did not show suppressor effect due to significant levels of soluble Sup35. About 15% of Sup35 was soluble, which, accounting for 20-fold overproduction, exceeded 3-fold the Sup35 levels in [psi^-] cells. Increased levels of soluble Sup35 indicate its inefficient polymerization, which may be related to the size of Sup35 polymers increased several-fold compared to conventional [PSI^+]. In cells with any given Sup35 levels an increase in polymer size would mean a proportional decrease in the number of Sup35 polymers and polymerization speed.

Another property of these polymers was poor heritability. Their appearance required the presence of [PIN^+] determinant. Furthermore, [PIN^+] elimination via disruption of *RNQ1* eliminated the amyloid-like Sup35 polymers in the cells which already have possessed them. Thus, in contrast to [PSI^+] polymers, which propagate very stably in the absence of [PIN^+], these polymers are not heritable. This may be related to small number of these polymers due to their inefficient fragmentation, which is evident from their large size. Then, the only reason for existence of these polymers is their efficient appearance de novo. This is likely to occur via seeding by Rnq1 polymers, since about 1/5 of total Rnq1 was bound to Sup35 polymers. This bond was resistant to cold SDS, and thus these Rnq1 and Sup35 belonged to the same polymers. It appears unlikely that Rnq1 was dispersed along the Sup35NM polymers, since homotypic polymerization should be much more efficient than heterotypic. More probably, this Rnq1 represented Rnq1 prion seeds attached to the Sup35NM polymers, which they initiated.

Finally, the nonheritable Sup35 amyloid-like polymers appear with very high frequency compared to its prion variants. A clear inverse correlation between the frequency of appearance and the "strength" was observed: stronger variants appeared less frequently.[50] The reason for this correlation is not fully clear. It may be presumed that this is due to the preferential survival of prion seeds corresponding to weaker [PSI^+], because they are less recognizable by the cellular chaperones. In this connection, it is noteworthy that among the Sup35 fibers spontaneously formed in vitro, a significant proportion apparently belongs to the prion type, since these fibers could transform yeast cells to the [PSI^+] phenotype.[26]

Thus, Sup35 forms amyloid variants covering the full spectrum of related phenotypes, from highly stable strong suppressor [*PSI*⁺] to nonheritable phenotypically undetectable polymers. The key parameter, distinguishing these variants is the frequency of polymer fragmentation, which is highest in strong [*PSI*⁺] polymers and lowest in nonheritable amyloid-like polymers. The ability of Sup35 to form both prions and nonheritable amyloids convincingly supports the idea that prion and amyloid phenomena are related. Notably, another amyloid-forming protein does not make full spectrum of variants: the Sup35 fusion protein with its prion-forming domain replaced with a stretch of 66 glutamine residues can form SDS-insoluble amyloid-like polymers, but is unable to form prions.[50]

Amyloid Cross-Seeding and the "Species Barrier"

Observations that the prion state of Sup35 may be seeded by polymers of Rnq1 and some other proteins lead to the suggestion that cross-seeding may play a role in the appearance of amyloids in mammals.[12] Recent work[50] showed that the Sup35 prion represents a very small fraction of all seeded Sup35 amyloid-like polymers, and thus the efficiency of cross-seeding is much higher than it was considered previously. This suggests that amyloid cross-seeding in mammals is not just possible, but plays a significant role in amyloid appearance. In agreement with this assumption, injection of synthetic amyloid fibers made of transthyretin or islet amyloid polypeptide caused deposition of amyloid A fibers.[51]

It is known that the transfer of PrP prion between different species occurs with difficulty or does not occur at all even in the cases when the inter-species difference in the sequence of prion proteins constitutes only few amino acids.[2] This effect, known as the "species barrier", is considered to result from inefficient copolymerization of differing prion proteins. The Rnq1-Sup35 pair provides a good model for this effect and uncovers an additional reason for it. In vitro, Rnq1 fibers seeded Sup35 polymerization, though about 100-fold less efficiently than did Sup35 fibers.[24] In vivo, such efficiency of seeding would be sufficient to cross the "species barrier", because, once appeared, the Sup35 prion will replicate independently. Yeast cell contains more than 1000 of Rnq1 molecules,[52] most of which polymerize per generation. Then, about 10 events of Sup35 seeding by Rnq1 may be expected per generation. However, as we mentioned, the majority of these events results in nonprion Sup35 amyloids.[50] Thus, the additional barrier for prion transmission is the loss of a specific prion fold, required for efficient fiber fragmentation. It should be noted that the sequence similarity between heterologous PrP proteins is much higher than the similarity between the prion domains of Sup35 and Rnq1. Therefore, in the PrP case the "polymerization" barrier may be lower, while the loss of fragmentable prion fold may become the main reason for prion "species barrier".

Prions and Nontransmissible Amyloids: Two Modes of the Polymerization Process

Prions of higher eukaryotes are infectious, while other amyloids are not. What are the reasons for this difference? The basic prerequisite of infectivity, the ability of polymers to catalyze further polymerization, is common for all amyloids. Another significant requirement is the ability of polymers to migrate between different organs and from one organism to another. The intracellular location of some amyloids (e.g., formed by α-Synuclein, Huntingtin) should restrict their infectivity, since amyloids, unlike viruses, do not have specific mechanisms for leaving and entering cells. However, the majority of amyloids are extracellular. A reason for noninfectivity may be the inability of consumed amyloid to pass the digestive tract and reach the appropriate organ. Apparently, in the PrP case these tasks are facilitated by its very high protease resistance and association with B-cells, which carry it around an organism.[53] Such specific properties are not modeled in yeast. However, usually the lack of infectivity is evident upon direct injection of amyloid material into appropriate tissue. This excludes these properties and allows considering the infectivity in terms of the factors acting in yeast.

For the yeast Sup35, we observed two modes of amyloid-like polymerization. In the "prion" mode, new amyloid particles appear from the existing ones by fragmentation. In the "nonprion amyloid" mode, new particles appear de novo. The fragmentation efficiency appears to represent the key difference between prion and nonprion amyloids in both yeast and mammals. The nonprion amyloid mode is noninfectious, because this process does not replicate old seeds, but generates new ones. The process does not depend on the introduction of external infection but depends on intrinsic propensity of an organism to generate and accumulate new amyloids. This propensity is known to increase with age, presumably due to age-dependent alterations of chaperone and/or protease systems. Consideration of physical properties of these amyloids also reveals reasons for their noninfectivity. Due to infrequent fragmentation, nonprion amyloid fibers are long and tend to precipitate. The commonly observed amyloid plaques represent an evident example of such behavior. If such amyloid polymers can not migrate to new locations, they can appear there only by formation de novo.

The mammalian PrP prion is distinguished by very low frequency of de novo appearance and should multiply via fragmentation. The animal fragmentation factor is not known, since *HSP104* homologues are not present in the sequenced genomes of man and invertebrates. Nevertheless, PrP prion particles appear to be fairly small in size,[54] which suggests the existence of efficient fragmentation factor. The lack of evident PrP deposits in many cases of Creutzfeldt-Jacob disease[2] may also be related to the small size of PrP polymers.

It is not clear why mammalian amyloidogenic proteins do not produce, like Sup35, efficiently fragmented polymer variants. One possible explanation is that the proteins with such ability would form polymers too easy and fast, thus being detrimental for an organism. Such properties should be counterselected by evolution and such proteins "fine tuned" to exclude prion formation by them. An example of such amyloid-only behavior in yeast is given by polyglutamine fusions to Sup35MC, which form only nonheritable polymers.[50] Another explanation could be insufficient fragmenting activity in the extracellular space. The efficient fragmentation of PrP may be related then to its specific life cycle, which includes both extracellular and intracellular phases.

The prion and amyloid polymerization modes represent two ideal extremes, while actual amyloids are likely to have intermediate properties, being fragmented, but infrequently. Intermediate properties may allow some amyloids to show infectivity under certain conditions. For example, polymerization of amyloid protein A, known as secondary systemic amyloidosis, may be induced by so-called amyloid enhancing factor, the active ingredient of which was identified as fibers of amyloid protein A.[55] However, the induction required additionally an inflammatory stimulus, such as silver nitrate. The mouse senile amyloidosis, related to polymerization of apolipoprotein A-II, showed properties principally similar to prions. This disease was transmitted by oral administration of apolipoprotein A-II fibers,[56] whose infectious potential was enhanced by ultrasonic fragmentation.

Acknowledgements

The work in the authors' laboratory was supported by the Wellcome Trust, Howard Hughes Medical Institute, International Science and Technology Center and Russian Foundation for Basic Research.

References

1. Horwich AL, Weissman JS. Deadly conformations — Protein misfolding in prion disease. Cell 1997; 89:499-510.
2. Prusiner SB, Scott MR, DeArmond SJ et al. Prion protein biology. Cell 1998; 93:337-348.
3. Koo EH, Lansbury Jr PT, Kelly JW. Amyloid diseases: Abnormal protein aggregation in neurodegeneration. Proc Natl Acad Sci USA 1999; 96:9989-9990.
4. Fowler DM, Koulov AV, Alory-Jost C et al. Functional amyloid formation within mammalian tissue. PLoS Biol 2005; 4:e6.
5. Si K, Lindquist S, Kandel ER. A neuronal isoform of the Aplysia CPEB has prion-like properties. Cell 2003; 115:879-891.

6. Wickner RB. [URE3] as an altered Ure2 protein: Evidence for a prion analog in Saccharomyces cerevisiae. Science 1994; 264:566-569.
7. Coustou V, Deleu C, Saupe S et al. The protein product of the het-s heterokaryon incompatibility gene of the fungus Podospora anserina behaves as a prion analog. Proc Natl Acad Sci USA 1997; 94:9773-9778.
8. Derkatch IL, Bradley ME, Zhou P et al. Genetic and environmental factors affecting the de novo appearance of the [PSI⁺] prion in Saccharomyces cerevisiae. Genetics 1997; 147:507-519.
9. Bradley ME, Edskes HK, Hong JY et al. Interactions among prions and prion "strains" in yeast. Proc Natl Acad Sci USA 2002; 99 Suppl 4:16392-16399.
10. Sondheimer N, Lindquist S. Rnq1: An epigenetic modifier of protein function in yeast. Mol Cell 2000; 5:163-172.
11. Santoso A, Chien P, Osherovich LZ et al. Molecular basis of a yeast prion species barrier. Cell 2000; 100:277-288.
12. Derkatch IL, Bradley ME, Hong JY et al. Prions affect the appearance of other prions: The story of [PIN⁺]. Cell 2001; 106:171-182.
13. Ter-Avanesyan MD, Dagkesamanskaya AR, Kushnirov VV et al. The SUP35 omnipotent suppressor gene is involved in the maintenance of the non-Mendelian determinant [psi⁺] in the yeast Saccharomyces cerevisiae. Genetics 1994; 137:671-676.
14. Derkatch IL, Chernoff YO, Kushnirov VV et al. Genesis and variability of [PSI⁺] prion factors in Saccharomyces cerevisiae. Genetics 1996; 144:1375-1386.
15. Liu JJ, Sondheimer N, Lindquist SL. Changes in the middle region of Sup35 profoundly alter the nature of epigenetic inheritance for the yeast prion [PSI⁺]. Proc Natl Acad Sci USA 2002; 99(Suppl 4):16446-16453.
16. Bradley ME, Liebman SW. The Sup35 domains required for maintenance of weak, strong or undifferentiated yeast [PSI⁺] prions. Mol Microbiol 2004; 51:1649-1659.
17. Ter-Avanesyan MD, Kushnirov VV, Dagkesamanskaya AR et al. Deletion analysis of the SUP35 gene of the yeast Saccharomyces cerevisiae reveals two nonoverlapping functional regions in the encoded protein. Mol Microbiol 1993; 7:683-692.
18. Zhou P, Derkatch IL, Uptain SM et al. The yeast non-Mendelian factor [ETA⁺] is a variant of [PSI⁺], a prion- like form of release factor eRF3. EMBO J 1999; 18:1182-1191.
19. Kochneva-Pervukhova NV, Chechenova MB, Valouev IA et al. [PSI ⁺] prion generation in yeast: Characterization of the "strain" difference. Yeast 2001; 18:489-497.
20. Coschigano PW, Magasanik B. The URE2 gene product of Saccharomyces cerevisiae plays an important role in the cellular response to the nitrogen source and has homology to glutathione S-transferases. Mol Cell Biol 1991; 11:822-832.
21. Patino MM, Liu JJ, Glover JR et al. Support for the prion hypothesis for inheritance of a phenotypic trait in yeast. Science 1996; 273:622-626.
22. Glover JR, Kowal AS, Schirmer EC et al. Self-seeded fibers formed by Sup35, the protein determinant of [PSI⁺], a heritable prion-like factor of S. cerevisiae. Cell 1997; 89:811-819.
23. King CY, Tittmann P, Gross H et al. Prion-inducing domain 2-114 of yeast Sup35 protein transforms in vitro into amyloid-like filaments. Proc Natl Acad Sci USA 1997; 94:6618-6622.
24. Derkatch IL, Uptain SM, Outeiro TF et al. Effects of Q/N-rich, polyQ, and nonpolyQ amyloids on the de novo formation of the [PSI+] prion in yeast and aggregation of Sup35 in vitro. Proc Natl Acad Sci USA 2004; 101:12934-12939.
25. King CY, Diaz-Avalos R. Protein-only transmission of three yeast prion strains. Nature 2004; 428:319-323.
26. Tanaka M, Chien P, Naber N et al. Conformational variations in an infectious protein determine prion strain differences. Nature 2004; 428:323-328.
27. Brachmann A, Baxa U, Wickner RB. Prion generation in vitro: Amyloid of Ure2p is infectious. EMBO J 2005; 24:3082-3092.
28. Kryndushkin DS, Alexandrov IM, Ter-Avanesyan MD et al. Yeast [PSI⁺] prion aggregates are formed by small Sup35 polymers fragmented by Hsp104. J Biol Chem 2003; 278:49636-49643.
29. Bagriantsev S, Liebman SW. Specificity of prion assembly in vivo. [PSI⁺] and [PIN⁺] form separate structures in yeast. J Biol Chem 2004; 279:51042-51048.
30. Jones GW, Tuite MF. Chaperoning prions: The cellular machinery for propagating an infectious protein? Bioessays 2005; 27:823-832.
31. Chernoff YO, Lindquist SL, Ono B et al. Role of the chaperone protein Hsp104 in propagation of the yeast prion- like factor [PSI⁺]. Science 1995; 268:880-884.
32. Moriyama H, Edskes HK, Wickner RB. [URE3] prion propagation in Saccharomyces cerevisiae: Requirement for chaperone Hsp104 and curing by overexpressed chaperone Ydj1p. Mol Cell Biol 2000; 20:8916-8922.

33. Parsell DA, Kowal AS, Singer MA et al. Protein disaggregation mediated by heat-shock protein Hsp104. Nature 1994; 372:475-478.
34. Glover JR, Lindquist S. Hsp104, Hsp70, and Hsp40: A novel chaperone system that rescues previously aggregated proteins. Cell 1998; 94:73-82.
35. Kushnirov VV, Ter-Avanesyan MD. Structure and replication of yeast prions. Cell 1998; 94:13-16.
36. Serio TR, Cashikar AG, Kowal AS et al. Nucleated conformational conversion and the replication of conformational information by a prion determinant. Science 2000; 289:1317-1321.
37. Ferreira PC, Ness F, Edwards SR et al. The elimination of the yeast [PSI$^+$] prion by guanidine hydrochloride is the result of Hsp104 inactivation. Mol Microbiol 2001; 40:1357-1369.
38. Tuite MF, Mundy CR, Cox BS. Agents that cause a high frequency of genetic change from [PSI$^+$] to [psi$^-$] in Saccharomyces cerevisiae. Genetics 1981; 98:691-711.
39. Eaglestone SS, Ruddock LW, Cox BS et al. Guanidine hydrochloride blocks a critical step in the propagation of the prion-like determinant [PSI$^+$] of Saccharomyces cerevisiae. Proc Natl Acad Sci USA 2000; 97:240-244.
40. Ness F, Ferreira P, Cox BS et al. Guanidine hydrochloride inhibits the generation of prion "seeds" but not prion protein aggregation in yeast. Mol Cell Biol 2002; 22:5593-5605.
41. Jung G, Jones G, Masison DC. Amino acid residue 184 of yeast Hsp104 chaperone is critical for prion-curing by guanidine, prion propagation, and thermotolerance. Proc Natl Acad Sci USA 2002; 99:9936-9941.
42. Kushnirov VV, Kochneva-Pervukhova NV, Chechenova MB et al. Prion properties of the Sup35 protein of yeast Pichia methanolica. EMBO J 2000; 19:324-331.
43. Schlumpberger M, Prusiner SB, Herskowitz I. Induction of distinct [URE3] yeast prion strains. Mol Cell Biol 2001; 21:7035-7046.
44. Kushnirov VV, Kryndushkin DS, Boguta M et al. Chaperones that cure yeast artificial [PSI$^+$] and their prion-specific effects. Curr Biol 2000; 10:1443-1446.
45. King CY. Supporting the structural basis of prion strains: Induction and identification of [PSI$^+$] variants. J Mol Biol 2001; 307:1247-1260.
46. Krishnan R, Lindquist SL. Structural insights into a yeast prion illuminate nucleation and strain diversity. Nature 2005; 435:765-772.
47. Tanaka M, Chien P, Yonekura K et al. Mechanism of cross-species prion transmission: An infectious conformation compatible with two highly divergent yeast prion proteins. Cell 2005; 121:49-62.
48. Song Y, Wu YX, Jung G et al. Role for Hsp70 chaperone in Saccharomyces cerevisiae prion seed replication. Eukaryot Cell 2005; 4:289-297.
49. Dagkesamanskaya AR, Ter-Avanesyan MD. Interaction of the yeast omnipotent suppressors SUP1(SUP45) and SUP2(SUP35) with non-Mendelian factors. Genetics 1991; 128:513-520.
50. Salnikova AB, Kryndushkin DS, Smirnov VN et al. Nonsense suppression in yeast cells overproducing Sup35 (eRF3) is caused by its nonheritable amyloids. J Biol Chem 2005; 280:8808-8812.
51. Johan K, Westermark G, Engstrom U et al. Acceleration of amyloid protein A amyloidosis by amyloid-like synthetic fibrils. Proc Natl Acad Sci USA 1998; 95:2558-2563.
52. Ghaemmaghami S, Huh WK, Bower K et al. Global analysis of protein expression in yeast. Nature 2003; 425:737-741.
53. Klein MA, Frigg R, Flechsig E et al. A crucial role for B cells in neuroinvasive scrapie. Nature 1997; 390:687-690.
54. Silveira JR, Raymond GJ, Hughson AG et al. The most infectious prion protein particles. Nature 2005; 437:257-261.
55. Lundmark K, Westermark GT, Nystrom S et al. Transmissibility of systemic amyloidosis by a prion-like mechanism. Proc Natl Acad Sci USA 2002; 99:6979-6984.
56. Xing Y, Nakamura A, Chiba T et al. Transmission of mouse senile amyloidosis. Lab Invest 2001; 81:493-499.

CHAPTER 7

Chaperone Effects on Prion and Nonprion Aggregates

Eugene G. Rikhvanov, Nina V. Romanova and Yury O. Chernoff*

Abstract

Exposure to high temperature or other stresses induces a synthesis of heat shock proteins. Many of these proteins are molecular chaperones, and some of them help cells to cope with heat-induced denaturation and aggregation of other proteins. In the last decade, chaperones have received increased attention in connection with their role in maintenance and propagation of the *Saccharomyces cerevisiae* prions, infectious or heritable agents transmitted at the protein level. Recent data suggest that functioning of the chaperones in reactivation of heat-damaged proteins and in propagation of prions is based on the same molecular mechanisms but may lead to different consequences depending on the type of aggregate. In both cases the concerted and balanced action of "chaperones' team", including Hsp104, Hsp70, Hsp40 and possibly other proteins, determines whether a misfolded protein is to be incorporated into an aggregate, rescued to the native state or targeted for degradation.

Chaperones and Thermotolerance

The Role of Hsps in Development of Induced Thermotolerance

A mild heat shock, which is nonlethal by itself, induces the synthesis of heat shock proteins (Hsps) and enhances the cell capacity to survive the subsequent severe heat shock exposure. This phenomenon is known as induced thermotolerance. It has been suggested that Hsp induction is essential for survival at elevated temperatures.[1] A correlation has been found between the dynamics of development of induced thermotolerance in *S. cerevisiae* and synthesis of the heat shock protein with the molecular mass of about 100 kDa, later designated as Hsp104.[2] Experiments with the deletion mutants clarified the role of Hsps in induced thermotolerance. Hsp104 is shown to play a crucial role in the induced thermotolerance in yeast,[1,3,4] while Hsps of the Hsp70,[4,5] Hsp40[4] and small Hsp[6,7] families perform auxiliary functions in this process. Our focus will be on the Hsp104, 70 and 40 which are shown to cooperate with each other in disaggregation and refolding of the stress-damaged proteins.

General Characteristics of Molecular Chaperones: "Holdases", "Foldases" and "Disaggregases"

Most Hsps are molecular chaperones facilitating protein folding, assembly and translocation across intracellular membranes. Roughly chaperones may be classified based on the mode of their interaction with substrate proteins. The ATP independent molecular chaperones, such

*Corresponding Author: Yury O. Chernoff—School of Biology, Georgia Institute of Technology, M/C 0230, 310 Ferst Drive, Atlanta, GA 30332-0230, U.S.A. Email: yury.chernoff@biology.gatech.edu

as small heat shock proteins (sHsps) and proteins of Hsp40 family that can stabilize unfolded polypeptides but could not reactivate them are referred to as "holdases".[6,8] Some chaperones having ATP-binding domains that assist in folding of nonnative proteins via ATP-dependent binding and release are known as "foldases". Proteins of the Hsp70 family also assist in folding, so that Hsp70s also seem to possess "foldase" activity. However, there is an alternative view considering the Hsp70 chaperones as "unfoldases" that use free energy from ATP binding and/or hydrolysis to unfold or pull apart misfolded and aggregated proteins to yield productive folding intermediates.[9]

The Hsp100/ClpB proteins are proposed to use ATP hydrolysis to disentangle aggregated polypeptides and transfer partially folded species to the Hsp70-Hsp40 binary system for subsequent refolding.[4,10] In this context, Hsp100s act as "disaggregases".[11] However, *Escherichia coli* homologues of Hsp70 (DnaK) and Hsp40 (DnaJ) are thought to bind first to aggregated proteins, potentially helping Hsp100/ClpB to extract polypeptides from aggregates.[10] Such an interplay is also shown to occur between sHsps and Hsp100/ClpB in yeast and bacteria.[6,7,10]

Yeast Hsp104

Hsp104 (Fig. 1A) is a yeast member of the Hsp100/CipB family of AAA (ATPases associated with various cellular activities) superfamily of proteins, participating in a variety of cellular activities.[10,11] In the presence of ATP, ADP or ATPγS, Hsp104 monomers are assembled into the hexamer complexes with an axial channel.[12] Yeast Hsp104 has two nucleotide-binding domains, NBD1 and NBD2 with different catalytic properties.[12,13] Mutations in NBD1 have little effect on hexamerization, while mutations in NBD2 severely impair hexamerization.[14] There is an allosteric communication between NBD1 and NBD2[13] as well as communication between individual monomers in a hexamer, so that the ATP hydrolysis by Hsp104 is greatly influenced by hexamerization.[15]

Hsp104 is induced by a mild heat shock treatment, and is crucial for induced thermotolerance in *S. cerevisiae*.[1] It is also known to be induced in response to hydrogen peroxide,[16] ethanol and sodium arsenite,[1] and near-freezing cold shock.[17] The ability of the *hsp104Δ* cells pretreated at 37°C to survive a lethal heat shock at 50°C is severely impaired but not completely abolished,

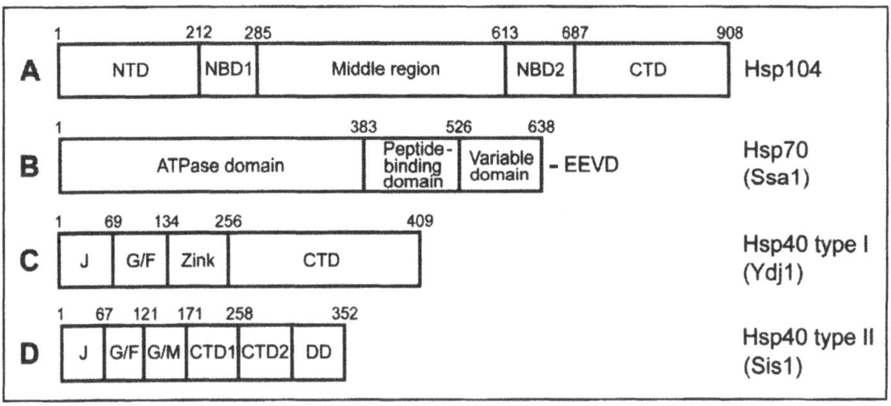

Figure 1. Structural organization of the yeast Hsp proteins involved in stress protection and prion propagation. A) Hsp104; B) Ssa1 as a representative of the Hsp70 family; C) Hsp40 type I (Ydj1); D) Hsp40 type II (Sis1). (The Hsp40 type III is not shown.) Designations: NBD, nucleotide-binding domain; NTD, N-terminal domain, middle region; CTD, C-terminal domain; J, J-domain; G/F, glycine and phenylalanine-rich region; Zinc, zinc-finger domain; G/M, glycine and methionine-rich region; DD, dimerization domain; E, glutamic acid; V, valine; D, aspartic acid. Numbers correspond to aa positions.

indicating that other heat induced proteins also play role in this process.[1] Specifically, Hsp70-Ssa and small heat shock protein Hsp26 become very important for thermotolerance in the *S. cerevisiae* cells lacking Hsp104.[5,7] Aggregation of cellular proteins is a major consequence of severe stress, and Hsp104 is thought to act directly on protein aggregates, leading to their resolubilization.[3] For efficient protein reactivation, Hsp104 requires the assistance of Hsp40 (Ydj1) and Hsp70 (Ssa).[4]

Abilities of Hsp104 to form a homohexamer and cooperatively bind and hydrolyze ATP are required for its functions in vivo.[12,13] When ATP binding to NBD1 is impaired by a mutation, Hsp104 is unable to interact with substrates both in vitro and in vivo.[18,19] Therefore, the ATP-bound state of NBD1 seems to be crucial for the chaperone-substrate interaction. Two models have been proposed for disaggregating action of Hsp104. In the first model, Hsp104 breaks up large aggregates into smaller ones in a crowbar-like activity. The second model suggests that a single polypeptide chain is extracted from an aggregate via translocation through the axial channel of Hsp104/ClpB hexamer, occurring by the unfolding/threading mechanism.[10,11] The structure of the pore entrance of the Hsp104 oligomeric complex was shown to be crucial for Hsp104 function,[20] supporting the latter mechanism.

Yeast Hsp70

All Hsp70 family proteins have three functionally separated domains: N-terminal 45 kDa ATP-binding domain, 15 kDa peptide-binding domain and C-terminal variable domain (Fig. 1B). Hsp70 transiently holds unfolded substrates in an intermediate state, preventing irreversible aggregation and catalyzing folding in the ATP dependent manner.[21] However, the precise mechanism by which Hsp70 promotes folding is unclear thus far. There are at least two possible models of Hsp70's action, which are not mutually exclusive. Model 1 ('kinetic partitioning' model) suggests that Hsp70 plays a rather passive role. Via repetitive substrate binding and release cycles, it decreases the concentration of a free substrate. This prevents aggregation and allows more time for the substrate to fold into the native state. Model 2 ('local unfolding' model) proposes that Hsp70 induces local unfolding in the substrate, e.g., the untangling of a misfolded β-sheet, which helps to overcome kinetic barriers for folding to the native state. ATP energy may be needed either to induce such conformational changes or alternatively, to drive the ATPase cycle in the right direction.[21] *S. cerevisiae* genome contains at least fourteen genes coding for the Hsp70 proteins. These proteins are localized in a variety of cellular compartments including the cytosol (subfamilies Ssa and Ssb), mitochondria (Ssc1 and Ssq1), endoplasmic reticulum (Kar2 and Lhs1), etc.[22]

Ssa subfamily is encoded by 4 genes, *SSA1, 2, 3* and *4* (reviewed in ref. 22). *SSA3* and *SSA4* genes are expressed only at very low level in the exponentially growing cells, but are drastically induced after the temperature upshift, as well as by the stationary phase and other stresses. *SSA2* is constitutively expressed, while *SSA1* is normally expressed at moderate levels and induced by stresses. Deletion of any individual *SSA* gene does not affect induced thermotolerance. Double *ssa1Δ ssa2Δ* mutants grew slower than the parent at all temperatures and were unable to form colonies at 37°C, but their ability to induce tolerance to heat shock at 37°C was not changed. Moreover, double mutant displayed a higher level basal thermotolerance that is apparently due to up-regulation of other Hsps.[22,23] At least one of the Ssa proteins must be present to preserve the cell viability.[22] Ssa proteins are implicated in protein translocation across intracellular membrane, prevention of aggregation of denatured proteins[24] and cotranslational folding.[25]

Another yeast cytosolic Hsp70 family, Ssb, is not stress-inducible, and is encoded by two almost identical genes, *SSB1* and *SSB2*. Strains with a single *SSB* gene disrupted exhibit no phenotypic change. However, a mutant with both genes disrupted grows slowly at all temperatures, and is cold sensitive.[22] The major fraction of Ssb proteins has been found in association with the translating ribosomes, although some Ssb is distributed freely in the yeast cytosol. It is postulated that Ssb aids in folding of the emerging newly synthesized proteins.[25] It is also possible that this protein is involved in protein turnover in the yeast by targeting misfolded proteins for degradation, as overexpression of Ssb1 suppressed the growth defects caused by some proteasome mutations.[26]

Yeast Hsp40 or J-Proteins

The Hsp40 family, subdivided into three subtypes, includes the structurally and functionally diverse proteins with one common feature, N-terminal J-domain (Fig. 1C,D).[9] According to the current model,[9,21] J-proteins first bind unfolded protein substrates in order to prevent their aggregation, and then transfer them to Hsp70, simultaneously stimulating the Hsp70 ATPase activity and thus stabilizing the Hsp70 interaction with the unfolded protein.

Type I Hsp40s, for instance yeast Ydj1, also contain the glycine and phenylalanine-rich (G/F) region, zinc finger-like domain, and conserved C-terminal domain (CTD). Type II Hsp40s, for example yeast Sis1, lack the zinc finger-like domain but contain extended glycine-rich region. The first 55 amino acids (aa) of this region are also rich in phenylalanines (G/F); the last 49 aa are rich in methionine residues (G/M). The C-terminal 181 aa of Sis1 contain the proposed polypeptide binding site (CTD1), a domain of unknown function (CTD2), and a dimerization domain.[9,27]

Both Ydj1 (type I) and Sis1 (type II) interact with Ssa, but not with Ssb. Ssb has its own Hsp40 cochaperone, zuotin or Zuo1.[9] Ydj1 is not essential, but *ydj1Δ* deletion causes severe growth defects. In cooperation with Ssa, Ydj1 promotes the protein translocation across the intracellular membranes, and participates in refolding of the heat-damaged proteins. C-terminal domain of Ydj1 has been implicated in binding unfolded polypeptides.[23] Apparently, zinc finger-like domain is necessary for transferring the nonnative polypeptides from Ydj1 to Hsp70.[28] Ydj1 is also required for ubiquitin-dependent degradation of certain abnormal proteins.[23]

The essential protein Sis1 is shown to be less effective than Ydj1 in helping Ssa to suppress aggregation of stress damaged proteins but is linked to other processes, for example initiation of protein synthesis.[25] Functions of Ydj1 and Sis1 are overlapping but not identical. Although excess Sis1 complements the slow growth phenotype of *ydj1Δ*, Ydj1 cannot complement the lethal phenotype of *sis1Δ*.[24]

Chaperone Effects on Prion Propagation

The Role of Hsp104 in Propagation of the Yeast Prions

Yeast prions [*PSI*⁺], [*PIN*⁺] and [*URE3*] are self-perpetuating amyloid-like polymers of the proteins Sup35, Rnq1 and Ure2, respectively (reviewed in refs. 29-31; for more details, see chapters 1,2 and 4-6). Role of chaperones in prion propagation was first demonstrated for [*PSI*⁺].[32] Search for genes that antagonize [*PSI*⁺] when present in the increased number of copies produced *HSP104*.[33] Further investigation revealed that both overproduction and inactivation of Hsp104 cause loss of [*PSI*⁺].[32] Surprisingly, overproduction of Hsp104 does not prevent de novo [*PSI*⁺] appearance.[34,35] Hsp104 is also required for propagation of [*PIN*⁺][36] and [*URE3*],[37] although high levels of Hsp104 do not antagonize these prions.

Recent experimental evidence[38-41] (reviewed in ref. 29) supports a model[35] postulating that the major role of Hsp104 in prion propagation in vivo is to break prion amyloids into smaller "seeds", initiating new rounds of prion production (Fig. 2). This mechanism, described in more details in chapters 5 and 6, essentially means that disaggregating activity of Hsp104 converts amyloids into self-perpetuating prions.[30,31] Variants of [*PSI*⁺] producing large aggregates that are relatively insensitive to Hsp104 require excess of this protein for efficient propagation.[35,39]

Dominant negative point mutations in either NBD (Fig. 1) disturb the Hsp104 ability to perform its role in both induced thermotolerance[3,4,12,13] and prion propagation,[13,32,38,42] indicating that both functions are associated with hexamerization and ATP hydrolysis. However, some mutant derivatives of Hsp104 that function efficiently in prion propagation can not protect yeast from extreme thermal stress.[19] On the other hand, the dominant mutant derivative of Hsp104, bearing the A503V substitution in the middle region, increases size of the Sup35 aggregates in [*PSI*⁺] cells, leading to accumulation of cytologically detectable clumps and cytotoxicity, but decreases aggregate size and cytotoxicity of the polyglutamine fragment of human huntingtin expressed in yeast, and does not affect thermotolerance.[43] A503V substitution impairs coordinated regulation of the NBD action without completely eliminating ATPase activity,[44] that possibly leads to different consequences depending on the type of aggregate that Hsp104 is interacting with.

It remains unknown why does excess Hsp104 cause loss of [*PSI*⁺] but not of the other yeast prions. [*PSI*⁺] loss in the presence of excess Hsp104 probably requires other Hsp104 activities in addition to (or instead of) those involved in prion propagation, as N-terminal region of Hsp104 is required for [*PSI*⁺] curing by excess Hsp104 but not for [*PSI*⁺] maintenance.[45] It is possible that excess Hsp104 solubilizes Sup35 prion polymers into monomers, that may require a mode of action distinct from one involved in oligomeric "seed" production. Alternatively, it was proposed that excess Hsp104 may impair prion segregation in cell divisions.[46] Indeed, average size of remaining Sup35 polymers is increased in the presence of excess Hsp104,[41] although this could be due to the fact that larger polymers are less sensitive to the Hsp104 disaggregating effect. The [*PSI*⁺] cells overexpressing Hsp104 accumulate cytologically detectable ring-like Sup35 structures that are not usually found in the [*PSI*⁺] cells with normal levels of Hsp104.[34] Rings are also observed in the cells undergoing de novo [*PSI*⁺] induction in the presence of excess Sup35,[34] and are shown to be associated with some components of the cortical actin cytoskeleton (CSK) involved in endocytosis, and/or with vacuolar membrane.[47] It is possible that rings represent intermediates that arise in the process of either de novo prion formation or prion elimination, and are attempted by the cell to be targeted for elimination via autophagy and vacuolar proteolysis.

Effects of Other Chaperones on Prions and Polyglutamine Aggregates in Yeast

Ssa Effects

Increased levels of Ssa1[48] or any other member of Ssa subfamily[49] enhanced phenotypic manifestation of [*PSI*⁺] and antagonized [*PSI*⁺] curing by excess Hsp104. Moreover, excess Ssa facilitated de novo [*PSI*⁺] induction in [*psi*⁻] cells by overproduced Sup35.[49] Several semi-dominant mutations in the *SSA1* gene have been obtained that decrease the mitotic stability of [*PSI*⁺].[50-52] Strains with the mutant alleles of *SSA1* were unable to propagate [*PSI*⁺] in the absence of the wild-type alleles of both *SSA* genes normally expressed in the absence of stress, *SSA1* and *SSA2*.[50] The second-site mutations in *SSA1* restored normal prion propagation.[51] Some (but not all) of the distantly related Hsp70 homologs from other organisms (plants and mammals) partly compensated the defect in [*PSI*⁺] maintenance observed in the presence of mutant Ssa.[45] This shows that Ssa functions involved in [*PSI*⁺] maintenance are at least to a certain extent conserved in evolution.

Both calculations based on kinetics of [*PSI*⁺] curing,[50] and biochemical assays such as size fractionation by chromatography and efficiency of fluorescence recovery after photobleaching indicated that size of the Sup35 aggregated structures is increased in the [*PSI*⁺] strains with mutant Ssa.[53] It has been interpreted as an evidence of that Ssa helps to disassemble large aggregated structure into smaller polymers that become a target for the "shearing" action of Hsp104. On the other hand, Ssa overproduction slightly but reproducibly increased average size of prion polymers produced from large structures in the semi-denaturing conditions and visualized by semi-denaturing gel electrophoresis, that was also accompanied by an increase in the proportion of monomeric Sup35.[49] As phenotypic manifestation of [*PSI*⁺] was enhanced rather than antagonized by excess Sup35, one could suggest that a significant fraction of the monomeric Sup35 generated in these cultures remained in the nonfunctional (possibly misfolded) state. Physical association between Ssa and Sup35 has been confirmed both in vivo and in vitro, suggesting that effects of Ssa on [*PSI*⁺] are due to its direct interaction with the prion-forming protein.[49]

While excess Ssa normally aids in propagation of the "conventional" variants of [*PSI*⁺], this effect could be reversed depending on the features of prion isolate and/or conditions affecting aggregate size and seed number. When Sup35 is overproduced in the [*PSI*⁺] strains, this leads to [*PSI*⁺] loss at low but detectable frequency, probably in result of accumulation of the large nontransmissible aggregates due to impairment of the balance between Sup35 and Hsp104.[49] Notably, this effect is exacerbated in the presence of excess Ssa. Likewise, excess Ssa antagonizes propagation of the [*PSI*⁺] derivatives that are characterized by the abnormally large aggregate size and require increased levels of Hsp104 for their propagation.[35,39]

In contrast to its effect on conventional [*PSI*⁺], overproduction of Hsp70-Ssa1 impaired propagation of the yeast prion [*URE3*].[54] Strangely enough, overproduction of the highly homologous Ssa2 protein did not show the same effect, and moreover, deletion of *SSA2* impaired propagation of [*URE3*].[55] Overproduction of some members of the Ssa subfamily counteracted poly-Q aggregation and/or toxicity in some yeast-based assays.[43,56] However, at least in one genotypic background poly-Q aggregation was also decreased by double *ssa1Δ ssa2Δ* deletion.[57]

Ssb Effects

In a strong contrast to Ssa, Ssb proteins consistently act as [*PSI*⁺] antagonists. Excess Ssb increases [*PSI*⁺] curing by Hsp104 overproduction,[58] inhibits [*PSI*⁺]-mediated suppression in certain [*PSI*⁺] isolates,[58] and causes loss of [*PSI*⁺] upon prolonged incubation in certain genotypic backgrounds.[59,60] Simultaneous deletion of both *SSB1* and *SSB2* genes decreases efficiency of [*PSI*⁺] curing by excess Hsp104 and increases the frequency of the spontaneous [*PSI*⁺] formation in [*psi*⁻] cells even in the absence of Sup35 overproduction.[58] Ssb, like Ssa, can directly interact with Sup35.[49] No effect of overproduced Ssb on poly-Q toxicity was detected.[43,56]

Difference between the Ssb and Ssa proteins in respect to [*PSI*⁺] curing is in significant part determined by their peptide-binding domains. The presence of Ssb peptide-binding domain is sufficient for an antiprion effect even when it is combined with the ATPase and variable domains of Ssa origin.[49]

Hsp40 Effects

Little is known about the effects of Hsp40 chaperones on yeast prions. Excess Ydj1 promoted loss of [*URE3*][37] and some variants of [*PIN*⁺],[61] and somewhat antagonized the chimeric prion [*PSI*⁺]$_{PS}$ generated by the Sup35 protein with a prion domain from the distantly related yeast *Pichia methanolica*.[62,63] The simultaneous overproduction of Ydj1 and Ssa1 cured some weak [*PSI*⁺] variants, but propagation of the strong variants remained unaffected.[62] On the other hand, Ydj1 deficiency did not affect maintenance of [*PIN*⁺][64] or [*PSI*⁺][51] in the absence of other chaperone mutations. In the cells carrying the semi-dominant mutation *SSA1-21*, lack of Ydj1 further impaired mitotic stability of [*PSI*⁺].[51]

Another yeast cochaperone of the Hsp40 family, Sis1, appears to be required for propagation of the [*PIN*⁺] prion.[64] Although Sis1 is an essential protein and therefore viable yeast cells lacking Sis1 cannot be constructed, in frame deletion within the *SIS1* gene eliminated [*PIN*⁺], although it did not affect [*PSI*⁺]. In contrast, overproduction of Sis1 had no detectable effect on [*PIN*⁺] propagation.[65] Sis1 is coprecipitated with the aggregated but not with the soluble form of Rnq1.[64] An extended glycine-rich region of Sis1, including a region rich in phenylalanine residues (G/F) is critical for prion maintenance.[64-66]

Data on the effects of Hsp40 proteins on poly-Q aggregation in yeast are somewhat contradictory. In some assays, overproduction of Ydj1 counteracted aggregation of some poly-Q constructs in yeast.[56] In the assay using prion-dependent poly-Q aggregation,[57] Ydj1 and Sis1 exhibited opposite effects: excess Ydj1 increased the size and toxicity of poly-Q aggregates generated in the [*PIN*$^+$] strain, while excess Sis1 decreased them.[42] Likewise, mutation in the *YDJ1* gene decreased poly-Q aggregation.[57] Interestingly, these effects of Ydj1 and Sis1 somewhat parallel at least some observations made with their human homologs (Hdj2 and Hdj1, respectively) in the mammalian cells.[67] Our preliminary data also indicate that overproduced Ydj1 and Sis1 proteins influence [*PSI*$^+$] curing by excess Hsp104 in the opposite ways (S. Müller, J. Patterson and Y. Chernoff, unpublished).

Model for the Chaperone Effects

Data reviewed above show that one and the same group of yeast chaperones is involved in protection against misfolded proteins and in prion propagation. We propose that effects of these chaperones on prion and nonprion aggregates are determined by one and the same molecular mechanism, while differences in effects are due to different parameters of aggregates. The model summarizing our current view is shown on Figure 2.

We propose that at normal levels, Hsp104 is responsible for propagation of the prion polymers via the subsequent cycles of breakage and growth. When the Hsp104/Sup35 ratio is shifted towards Sup35 in the [*PSI*$^+$]-containing cells, polymer size is increased leading to the accumulation of large clump-shaped inclusion bodies (IBs) that are eliminated from the population either due to a segregation defect, followed by the death of IB-accumulating cells, or via autophagy. Abundance of such IBs is increased in the [*PSI*$^+$] variants with the decreased sensitivity to Hsp104 (such as [*PSI*$^+$]$^{\uparrow 104d}$)[35] resulting in frequent loss of these variants in the normal conditions and their rescue at high levels of Hsp104.

Increase in the Hsp104 levels results in disruption of the ordered structure of prion aggregates. However, if overexpression of Hsp104 is not accompanied by the overexpression of the Hsp70 and Hsp40 chaperones, Hsp104 is not capable of solubilizing aggregates into the properly refolded monomers on its own. Therefore, disruption of the aggregate structure and increased hydrophobic exposure, induced by Hsp104, are followed by amorphous agglomeration of the misfolded prion protein molecules, rather than by solubilization. We propose that these misfolded agglomerated proteins can be recognized by Sis1, which in combination with Ssa targets them for either refolding or degradation via the proteasomal pathway.

Alternatively, agglomerates are recognized by the CSK networks involved in endocytosis, including such proteins as Sla1, Sla2 and End3.[47] This results in formation of the different

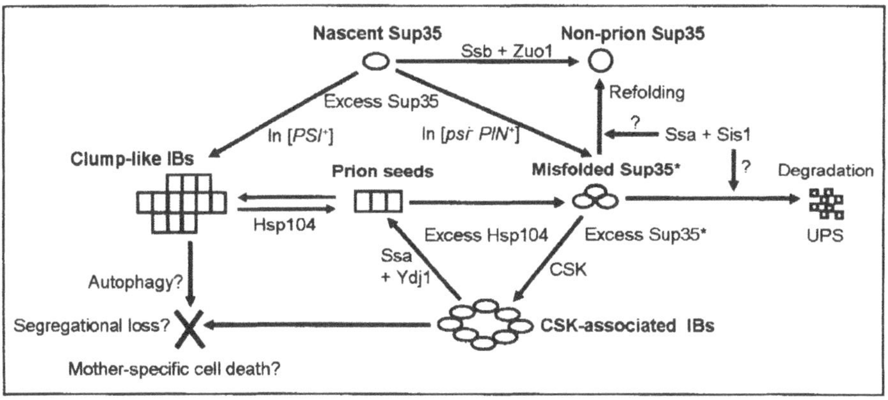

Figure 2. Model for the role of molecular chaperones in formation and propagation of the [*PSI*$^+$] prion. CSK, cytoskeleton; UPS, ubiquitin-proteasome system; IBs, inclusion bodies. See comments in text.

kind of IBs, frequently having the ring-shaped morphology and associated with CSK. These IBs are intended to be targeted for degradation via autophagy and vacuolar pathway, but they can also be recognized by Ydj1, which in combination with Ssa stabilizes proteins in the misfolded state and promotes disassembly of IBs to oligomers, capable of reentering the prion propagation cycle. Therefore, Ssa effects depend on the participating cochaperones.

The Ssb protein is acting on the nascent Sup35 polypeptide, and its major function is to promote folding of the newly synthesized Sup35 into a nonprion form. In this way, Ssb is antagonizing prion formation and propagation, and its effects are additive to all other anti-prion factors as it is working in a separate pathway.

The same model also explains chaperone effects on de novo [*PSI*$^+$] formation, as shown on Figure 2. Some parts of this model should certainly apply to the other prions as well, although other steps could turn out to be protein specific. E.g., effects of specific Hsps may vary depending on the prion, for example due to different affinities of the Hsp40s to different prion proteins. In case of polyglutamine aggregates lacking the prion propagating activity on their own, prevention of the aggregate degradation in the presence of excess Ydj1 leads to an increase in their size and toxicity.

Acknowledgements

This work was supported by grants R01GM58763 from NIH to YOC and 04-04-48275a from Russian Foundation of Basic Research to EGR.

References

1. Parsell DA, Lindquist S. The function of heat-shock proteins in stress tolerance: Degradation and reactivation of damaged proteins. Annu Rev Genet 1993; 27:437-496.
2. McAlister L, Finkelstein DB. Heat shock proteins and thermal resistance in yeast. Biochem Biophys Res Commun 1980; 93:819-824.
3. Parsell DA, Kowal AS, Singer M et al. Protein disaggregation mediated by heat-shock protein Hsp104. Nature 1994; 372:475-478.
4. Glover JR, Lindquist S. Hsp104, Hsp70 and Hsp40: A novel chaperone system that rescues previously aggregated proteins. Cell 1998; 94:73-82.
5. Sanchez Y, Parsell DA, Taulien J et al. Genetic evidence for a functional relationship between Hsp104 and Hsp70. J Bacteriol 1993; 175:6484-6491.
6. Haslbeck M, Franzmann T, Weinfurtner D et al. Some like it hot: The structure and function of small heat-shock proteins. Nat Struct Mol Biol 2005; 12:842-846.
7. Cashikar AG, Duennwald ML, Lindquist SL. A chaperone pathway in protein disaggregation: Hsp26 alters the nature of protein aggregates to facilitate reactivation by Hsp104. J Biol Chem 2005; 280:23869-23875.
8. Slepenkov SV, Witt SN. The unfolding story of the Escherichia coli Hsp70 DnaK: Is DnaK a holdase or an unfoldase? Mol Microbiol 2002; 45:1197-1206.
9. Walsh P, Bursac D, Law YC et al. The J-protein family: Modulating protein assembly, disassembly and translocation. EMBO Rep 2004; 5:567-571.
10. Weibezahn J, Schlieker C, Tessarz P et al. B. Novel insights into the mechanism of chaperone-assisted protein disaggregation. Biol Chem 2005; 386(8):739-744.
11. Bosl B, Grimminger V, Walter S. The molecular chaperone Hsp104 - A molecular machine for protein disaggregation. J Struct Biol 2006; 156:139-148.
12. Parsell DA, Kowal AS, Lindquist SL. Saccharomyces cerevisiae Hsp104 protein. Purification and characterization of ATP-induced structural changes. J Biol Chem 1994; 269:4480-4487.
13. Hattendorf DA, Lindquist SL. Cooperative kinetics of both Hsp104 ATPase domains and interdomain communication revealed by AAA sensor-1 mutants. EMBO J 2002; 21:12-21.
14. Schirmer EC, Queitsch C, Kowal AS et al. The ATPase activity of Hsp104, effects of environmental conditions and mutations. J Biol Chem 1998; 273:15546-15552.
15. Schirmer EC, Ware DM, Queitsch C et al. Subunit interactions influence the biochemical and biological properties of Hsp104. Proc Natl Acad Sci USA 2001; 98:914-919.
16. Godon C, Lagniel G, Lee J et al. The H_2O_2 stimulon in Saccharomyces cerevisiae. J Biol Chem 1998; 273:22480-22489.
17. Kandror O, Bretschneider N, Kreydin E et al. Yeast adapt to near-freezing temperatures by STRE/Msn2,4-dependent induction of trehalose synthesis and certain molecular chaperones. Mol Cell 2004; 13:771-781.

18. Bosl B, Grimminger V, Walter S. Substrate binding to the molecular chaperone Hsp104 and its regulation by nucleotides. J Biol Chem 2005; 280:38170-38176.
19. Tkach JM, Glover JR. Amino acid substitutions in the C-terminal AAA+ module of Hsp104 prevent substrate recognition by disrupting oligomerization and cause high temperature inactivation. J Biol Chem 2004; 279:35692-35701.
20. Lum R, Tkach JM, Vierling E et al. Evidence for an unfolding/threading mechanism for protein disaggregation by Saccharomyces cerevisiae Hsp104. J Biol Chem 2004; 279:29139-29146.
21. Mayer MP, Bukau B. Hsp70 chaperones: Cellular functions and molecular mechanism. Cell Mol Life Sci 2005; 62:670-684.
22. Craig EA, Gambill BD, Nelson RJ. Heat shock proteins: Molecular chaperones of protein biogenesis. Microbiol Rev 1993; 57:402-414.
23. Nelson RJ, Heschl MF, Craig EA. Isolation and characterization of extragenic suppressors of mutations in the SSA hsp70 genes of Saccharomyces cerevisiae. Genetics 1992; 131:277-85.
24. Fan CY, Lee S, Cyr DM. Mechanisms for regulation of Hsp70 function by Hsp40. Cell Stress Chaperones 2003; 8:309-316.
25. Craig EA, Eisenman HC, Hundley HA. Ribosome-tethered molecular chaperones: The first line of defense against protein misfolding? Curr Opin Microbiol 2003; 6:157-162.
26. Ohba M. Modulation of intracellular protein degradation by SSB1-SIS1 chaperon system in yeast S. cerevisiae. FEBS Lett 1997; 409:307-311.
27. Sha B, Lee S, Cyr DM. The crystal structure of the peptide-binding fragment from the yeast Hsp40 protein Sis1. Sructure 2000; 8:799-807.
28. Fan CY, Ren HY, Lee P et al. The type I Hsp40 zinc finger-like region is required for Hsp70 to capture nonnative polypeptides from Ydj1. J Biol Chem 2005; 280:695-702.
29. Chernoff YO. Mutation processes at the protein level: Is Lamarck back? Mutat Res 2001; 488:39-64.
30. Chernoff YO. Replication vehicles of protein-based inheritance. Trends Biotechnol 2004; 22:549-552.
31. Chernoff YO. Amyloidogenic domains, prions and structural inheritance: Rudiments of early life or recent acquisition? Curr Opin Chem Biol 2004; 8:665-671.
32. Chernoff YO, Lindquist SL, Ono B et al. Role of the chaperone protein Hsp104 in propagation of the yeast prion-like factor [psi+]. Science 1995; 268:880-884.
33. Chernoff YO, Ono B. Dosage-dependent modifiers of psi-dependent omnipotent suppression in yeast. In: Brown AJP, ed. Protein Synthesis and Targeting in Yeast. NATO ASI, 1993:H71:101-110.
34. Zhou P, Derkatch IL, Liebman SW. The relationship between visible intracellular aggregates that appear after overexpression of Sup35 and the yeast prion-like elements [PSI+] and [PIN+]. Mol Microbiol 2001; 39:37-46.
35. Borschenius AS, Muller S, Newnam GP et al. Prion variant maintained only at high levels of the Hsp104 disaggregase. Curr Genet 2006; 49:21-29.
36. Derkatch IL, Bradley ME, Zhou P et al. Genetic and environmental factors affecting the de novo appearance of the [PSI+] prion in Saccharomyces cerevisiae. Genetics 1997; 147:507-519.
37. Moriyama H, Edskes HK, Wickner RB. [URE3] prion propagation in Saccharomyces cerevisiae: Requirement for chaperone Hsp104 and curing by overexpressed chaperone Ydj1p. Mol Cell Biol 2000; 20:8916-8922.
38. Wegrzyn RD, Bapat K, Newnam GP et al. Mechanism of prion loss after Hsp104 inactivation in yeast. Mol Cell Biol 2001; 21:4656-4669.
39. Borchsenius AS, Wegrzyn RD, Newnam GP et al. Yeast prion protein derivative defective in aggregate shearing and production of new 'seeds'. EMBO J 2001; 20:6683-6691.
40. Ness F, Ferreira P, Cox BS et al. Guanidine hydrochloride inhibits the generation of prion "seeds" but not prion protein aggregation in yeast. Mol Cell Biol 2002; 22:5593-5605.
41. Kryndushkin DS, Alexandrov IM, Ter-Avanesyan MD et al. Yeast [PSI+] prion aggregates are formed by small Sup35 polymers fragmented by Hsp104. J Biol Chem 2003; 278(49):49636-49643.
42. Patino MM, Liu JJ, Glover JR et al. Support for the prion hypothesis for inheritance of a phenotypic trait in yeast. Science 1996; 273:622-626.
43. Gokhale KC, Newnam GP, Sherman MY et al. Modulation of prion-dependent polyglutamine aggregation and toxicity by chaperone proteins in the yeast model. J Biol Chem 2005; 280:22809-22818.
44. Schirmer EC, Homann OR, Kowal AS et al. Dominant gain-of-function mutations in Hsp104p reveal crucial roles for the middle region. Mol Biol Cell 2004; 15:2061-2072.
45. Hung GC, Masison DC. N-terminal domain of yeast Hsp104 chaperone is dispensable for thermotolerance and prion propagation but necessary for curing prions by Hsp104 overexpression. Genetics 2006; 173:611-20.
46. Cox B, Ness F, Tuite M. Analysis of the generation and segregation of propagons: Entities that propagate the [PSI+] prion in yeast. Genetics 2003; 165:23-33.

47. Ganusova EE, Ozolins LN, Bhagat S et al. Modulation of prion formation, aggregation, and toxicity by the actin cytoskeleton in yeast. Mol Cell Biol 2006; 26:617-629.
48. Newnam GP, Wegrzyn RD, Lindquist SL et al. Antagonistic interactions between yeast chaperones Hsp104 and Hsp70 in prion curing. Mol Cell Biol 1999; 19:1325-1333.
49. Allen KD, Wegrzyn RD, Chernova TA et al. Hsp70 chaperones as modulators of prion life cycle: Novel effects of Ssa and Ssb on the Saccharomyces cerevisiae prion [PSI+]. Genetics 2005; 169:1227-1242.
50. Jung G, Jones G, Wegrzyn RD et al. A role for cytosolic Hsp70 in yeast [PSI+] prion propagation and [PSI+] as a cellular stress. Genetics 2000; 156:559-570.
51. Jones GW, Masison DC. Saccharomyces cerevisiae Hsp70 mutations affect [PSI+] prion propagation and cell growth differently and implicate Hsp40 and tetratricopeptide repeat cochaperones in impairment of [PSI+]. Genetics 2003; 163:495-506.
52. Jones G, Song Y, Chung S et al. Propagation of Saccharomyces cerevisiae [PSI+] prion is impaired by factors that regulate Hsp70 substrate binding. Mol Cell Biol 2004; 24:3928-3937.
53. Song Y, Wu YX, Jung G et al. Role for Hsp70 chaperone in Saccharomyces cerevisiae prion seed replication. Eukaryot Cell 2005; 4:289-297.
54. Schwimmer C, Masison DC. Antagonistic interactions between yeast [PSI+] and [URE3] prions and curing of [URE3] by Hsp70 protein chaperone Ssa1p but not by Ssa2p. Mol Cell Biol 2002; 22:3590-3598.
55. Roberts BT, Moriyama H, Wickner RB. [URE3] prion propagation is abolished by a mutation of the primary cytosolic Hsp70 of budding yeast. Yeast 2004; 21(2):107-117.
56. Muchowski PJ, Schaffar G, Sittler A et al. Hsp70 and Hsp40 chaperones can inhibit self-assembly of polyglutamine proteins into amyloid-like fibrils. Proc Natl Acad Sci USA 2000; 97:7841-7846.
57. Meriin AB, Zhang X, He X et al. Huntington toxicity in yeast model depends on polyglutamine aggregation mediated by a prion-like protein Rnq1. J Cell Biol 2002; 157:997-1004.
58. Chernoff YO, Newnam GP, Kumar J et al. Evidence for a protein mutator in yeast: Role of the Hsp70-related chaperone Ssb in formation, stability, and toxicity of the [PSI] prion. Mol Cell Biol 1999; 19:8103-8112.
59. Kushnirov VV, Kryndushkin DS, Boguta M et al. Chaperones that cure yeast artificial [PSI+] and their prion-specific effects. Curr Biol 2000; 10:1443-1446.
60. Chacinska A, Szczesniak B, Kochneva-Pervukhova NV et al. Ssb1 chaperone is a [PSI+] prion-curing factor. Curr Genet 2001; 39:62-67.
61. Bradley ME, Edskes HK, Hong JY et al. Interactions among prions and prion "strains" in yeast. Proc Natl Acad Sci USA 2002; 99:16392-16399.
62. Kushnirov VV, Kochneva-Pervukhova NV, Chechenova MB et al. Prion properties of the Sup35 protein of yeast Pichia methanolica. EMBO J 2000; 19:324-331.
63. Kryndushkin DS, Smirnov VN, Ter-Avanesyan MD et al. Increased expression of Hsp40 chaperones, transcriptional factors, and ribosomal protein Rpp0 can cure yeast prions. J Biol Chem 2002; 277:23702-23708.
64. Sondheimer N, Lopez N, Craig EA et al. The role of Sis1 in the maintenance of the [RNQ+] prion. EMBO J 2001; 20:2435-2442.
65. Aron R, Lopez N, Walter W et al. In vivo bipartite interaction between the Hsp40 Sis1 and Hsp70 in Saccharomyces cerevisiae. Genetics 2005; 169:1873-1882.
66. Lopez N, Aron R, Craig EA. Specificity of class II Hsp40 Sis1 in maintenance of yeast prion [RNQ+]. Mol Biol Cell 2003; 14:1172-1181.
67. Wyttenbach A, Carmichael J, Swartz J et al. Effects of heat shock, heat shock protein 40 (HDJ-2), and proteasome inhibition on protein aggregation in cellular models of Huntington's disease. Proc Natl Acad Sci USA 2000; 97:2898-2903.

CHAPTER 8

Biological Roles of Prion Domains

Sergey G. Inge-Vechtomov, Galina A. Zhouravleva and Yury O. Chernoff*

Abstract

In vivo amyloid formation is a widespread phenomenon in eukaryotes. Self-perpetuating amyloids provide a basis for the infectious or heritable protein isoforms (prions). At least for some proteins, amyloid-forming potential is conserved in evolution despite divergence of the amino acid (aa) sequences. In some cases, prion formation certainly represents a pathological process leading to a disease. However, there are several scenarios in which prions and other amyloids or amyloid-like aggregates are either shown or suspected to perform positive biological functions. Proven examples include self/nonself recognition, stress defense and scaffolding of other (functional) polymers. The role of prion-like phenomena in memory has been hypothesized. As an additional mechanism of heritable change, prion formation may in principle contribute to heritable variability at the population level. Moreover, it is possible that amyloid-based prions represent by-products of the transient feedback regulatory circuits, as normal cellular function of at least some prion proteins is decreased in the prion state.

Introduction: Prions as the Second Order Templates

The central dogma of molecular biology[1] provides a specific mechanism for the previously postulated[2] template principle in biology. DNA and RNA can be considered as the first order templates, that is, linear or sequence templates, either for each other or for polypeptides. Discovery of infectious proteins (prions),[3] and especially of prion mechanism of inheritance[4] introduced templates of another type, structural or conformational templates, which could be designated as second order templates.[5] According to the current view,[4,6,7] the process of propagation of the amyloid-based prions begins with a conformational change in the protein, and is followed by "linear crystallization", producing amyloid fibers. The new rounds of prion multiplication may be initiated or seeded with preexisting amyloid fragments. Transmission of these fragments in cell divisions results in the inheritance of the prion state in yeast and fungal systems. Physiochemical studies of elementary amyloid particles uncovered the ß-rich structure,[8-10] in some examples[11] held together by the intermolecular parallel β-sheets. Variations of this structure apparently determine patterns of the specific variants, or "strains" of a given prion protein[12] (for more detail, see chapters 1 and 2). Conformation templating of a yeast prion can be reproduced in vitro,[13,14] resulting in generation of infectious prion particles, faithfully reproducing the variant-specific patterns upon transformation into the yeast cells. Yeast and fungal prions known to date are described in detail in other chapters of this book. Patterns of the mammalian prion protein PrP have been reviewed recently.[15,16]

Prion propagation is a highly sequence-specific process. Domains forming an axis of the amyloid fiber should be identical to each other at the level comparable to that required for the complementary interaction of nucleic acid sequences. However, aggregating proteins of

*Corresponding Author: Yury O. Chernoff—School of Biology, Georgia Institute of Technology, M/C 0230, 310 Ferst Drive, Atlanta, Georgia 30332-0230, U.S.A. Email: yury.chernoff@biology.gatech.edu

Protein-Based Inheritance, edited by Yury O. Chernoff. ©2007 Landes Bioscience.

different sequences can facilitate aggregation of each other in certain assays. For example, de novo appearance of the prion conformation of the yeast protein Sup35, containing a QN-rich prion domain, is facilitated in the presence of the prion isoform of another QN-rich protein, Rnq1,[17-19] reflecting the existence of a prion network. Molecular mechanism of this interaction is still unclear, and it does not seem to involve a template-like component (for more detail, see chapter 4).

Amyloid is probably an ancient fold, as almost any protein can form an amyloid in vitro depending on conditions.[20,21] Moreover, second order templating is not restricted to prions. Other examples of "structural inheritance" involve inheritance of preformed structures in Protozoa (reviewed in chapter 9).

As prions and similar phenomena appear to be widespread, the question arises whether these phenomena play a biological role. Two possible models of the biological role of prions were proposed in literature. One model, designated here and further as "prion pathology" model, states that prion (or amyloid) formation is a pathological process, while conservation of amyloid-forming potential in evolution is due to other adaptive functions of prion-forming proteins, which are not necessarily related to prion formation per se (e. g., see ref. 22). Another model, designated here and further as model of "adaptive prionization", suggests that prion formation by itself could be an adaptive process, so that certain prions are responsible for adaptive traits (e. g., see ref. 23).

"Prion Pathology" Model

Mammalian Prion Diseases and Other Aggregation-Related Diseases

Examples of the "prion diseases" are well known and include various infectious neurodegenerative diseases in mammals.[15,16] According to the "protein only" concept, which is now accepted by the majority of experts, the PrP protein in its prion form (PrP^{Sc}) is the sole component of a "transmissible particle" that is responsible for the genesis and transmission of a disease. Usually, there is a correlation between the disease and cerebral accumulation of PrP.[3,24] The properties of PrP are very similar to those seen in various noninfectious amyloidoses and neural inclusion disorders, a large and heterogeneous group including more than 20 human diseases, among them Alzheimer's, Huntington's and Parkinson's diseases,[25] resulting from conversion of certain proteins or their fragments from the normally soluble form to insoluble fibrils or plaques.

Although protein-destabilizing mutations can confer the ability to form amyloids in vivo even to such commonly known proteins as lysozyme,[26] usually disease-related aggregation depends on the presence of the specific elements of the primary structure. One feature frequently associated with aggregation is the presence of regions within proteins that comprise a single homopolymeric tract of a particular amino acid and are called homopeptide repeats, or SSR (single sequence repeats).[27] It has been shown that uncontrolled genetic expansions of SSR regions lead to the development of some neurodegenerative disorders, for example Huntington's disease, associated with the expanded poly-Q tract in the protein called huntingtin.[28] Several other diseases involve different proteins with poly-Q tracts but exhibit a similar mechanism of pathology. It was also demonstrated that some SSRs not linked to the specific disease are toxic to cells when overexpressed and/or lead to protein aggregation.[29-31]

These and the other facts indicate that accumulation of the amyloid-like aggregates is a pathological process. This notion is further confirmed by the existence of mechanisms preventing amyloid-like protein aggregation, such as a specific chaperone preventing aggregation of excess α-globin chains.[32] As misfolded and potentially aggregating proteins are usually accumulated during aging, it is an intriguing possibility that aging could promote prion-like pathologies. Indeed, some aggregation-related diseases (e.g., Alzheimer's disease) in humans are frequently associated with advanced age.

Exact mechanism of cell death in amyloid and neural inclusion disorders remains unknown. At least in case of mammalian PrP, it is certainly not due to lack of the normal protein function, as deletion of the PrP-coding gene does not cause a disease in mice.[33] For Huntington's disease, it is proposed that aggregates sequester some essential cellular proteins.[34-37] Poly-Q constructs introduced into *Caenorhabditis elegans* induce heat shock response at the stringency proportional to the length of the poly-Q stretch,[38] and disrupt the global quality control of protein folding, possibly by interfering with the disposal of misfolded proteins.[34] There is evidence that PrP and some other amyloidogenic proteins trigger cell death via apoptosis or autophagy.[39-42]

Pathological Effects of Amyloids in Yeast

Aggregated (prion) forms of the yeast proteins Sup35 and Ure2, called respectively [*PSI*⁺] and [URE3], are not found in the natural, industrial and clinical isolates of *Saccharomyces* yeast[22,43,44] (see also chapter 1 for more detail), consistent with the possibility of their pathogenicity. However the prion form of Rnq1 protein, called [*PIN*⁺], was found in a few natural isolates.[22,44] [URE3] decreases the growth rate of yeast.[22] While [*PSI*⁺] does not affect growth rates of exponential cells,[45] some [*PSI*⁺] strains exhibit facilitated cell death in the deep stationary phase, similar to apoptotic processes in higher eukaryotes (Y. Chernoff, J. Kumar, and G. Newnam, unpublished data). Activation of the apoptosis-like programmed cell death pathways in the starving yeast cells has been reported previously.[46-48] Some combinations of [*PSI*⁺] and [URE3] isolates exhibit the synthetic lethal or sublethal interactions.[49]

Overproduced Sup35 protein or fragments containing the Sup35 prion domain (Sup35N) are toxic to the [*PSI*⁺] cells, or (at very high levels) to the [*psi*⁻] cells containing the [*PIN*⁺] prion, that facilitates de novo [*PSI*⁺] induction.[17,50-52] This toxicity is not simply due to accumulation of excess protein per se, as it is controlled by the same protein regions that are involved in prion formation, and is not seen in the [*psi*⁻ *pin*⁻] background.[17,52] It is shown that accumulation of aggregated Sup35 in the prion-containing cells is associated with cell death.[53,54] This somewhat parallels mammalian prion diseases, where PrP^{Sc}-related pathology is usually detected only in neurons, cells known to produce mammalian prion protein (PrP) at high levels.[3]

Some mammalian amyloidogenic proteins are also toxic to yeast. The poly-Q expanded fragment of human huntingtin, fused to the green fluorescent protein (GFP) generates aggregates and causes toxicity only in yeast cells containing the endogeneous QN-rich prions, [*PIN*⁺][55] or [*PSI*⁺],[56] which manifest themselves as susceptibility factors for a poly-Q disorder. Prion-dependent poly-Q cytotoxicity in yeast is associated with a defect of endocytosis, apparently due to sequestration of some actin-assembly proteins, involved in formation of the endocytic vesicles, by poly-Q aggregates.[57] Sup35 aggregates also interact with some cytoskeletal proteins involved in the endocytic/vacuolar pathway, and cytotoxicity of overproduced Sup35 is increased in the strains with the cytoskeletal defects.[53,58] Expression of mammalian α-synuclein in yeast leads to its aggregation and cytotoxicity with some characteristics of apoptosis.[59] Taken together, these data confirm that accumulation of prions and other amyloidogenic protein in yeast may lead to the pathological consequences, and establish yeast prions as appropriate models for studying the mechanisms of amyloid cytotoxicity.

Model of "Adaptive Prionization"

Evolutionary Conservation of Prion-Forming Properties

One argument in favor of the adaptive role of prions is evolutionary conservation of prion-forming properties of some proteins. Prion properties of Sup35 are conserved in *Candida albicans* and *Pichia methanolica*, budding yeast species that are distantly related to *Saccharomyces cerevisiae*.[43,60-63] Comparison of the Sup35 sequences among the different isolates of *S. cerevisiae* and between the sister species of *S. cerevisiae* and *S. paradoxus* demonstrates that while the prion domain (Sup35N) is evolving much faster than the C-proximal release factor domain

(Sup35C), sequence of Sup35N still remains under the purifying selection pressure, confirming that this region of the protein is playing a certain positive biological role.[64] As the ability to form a prion is the only function of Sup35N known thus far, the simplest logical explanation would be that the ability to form a prion is adaptive under certain circumstances. Remarkably, the highest level of sequence conservation was observed within two subregions of Sup35N, the N-proximal QN-rich stretch (QN) and the region of oligopeptide repeats (ORs, see Fig. 1), which are still clearly seen in the distantly related budding yeast species of *Candida* and *Pichia*, despite low overall conservation of the Sup35N aa sequence (reviewed in ref. 65). Both subregions play a major role in prion-related properties of Sup35N (reviewed in ref. 7). However, these observations can argue in both ways, as repetitive structure of OR region per se is not a requirement for prion propagation.[66] Then, conservation of OR region (and possibly of QN) could be related to some unknown function of this part of the protein that is distinct from its prion-propagating ability.

The Sup35N region of the distant relative of budding yeast, the fission yeast *Schizosaccharomyces pombe*, does not contain QN and ORs (Fig. 2) and exhibits essentially no aa identity (only 18%) with the corresponding domain of *S. cerevisiae*, while Sup35C remains highly conserved (64% identity).[65] Likewise, neither sequence nor aa composition patterns of Sup35N are conserved between yeast and mammals, and the capability of Sup35 homologs (usually called eRF3) from species other than budding yeast to form prions is yet to be proven (Fig. 2). However, while aa composition of the Sup35N regions of higher eukaryotes is different from yeast Sup35N, it is still highly unusual. E.g., N-terminal domain of the Sup35 homolog from mouse and human (GSPT1) contain a high percentage of P, S and G residues

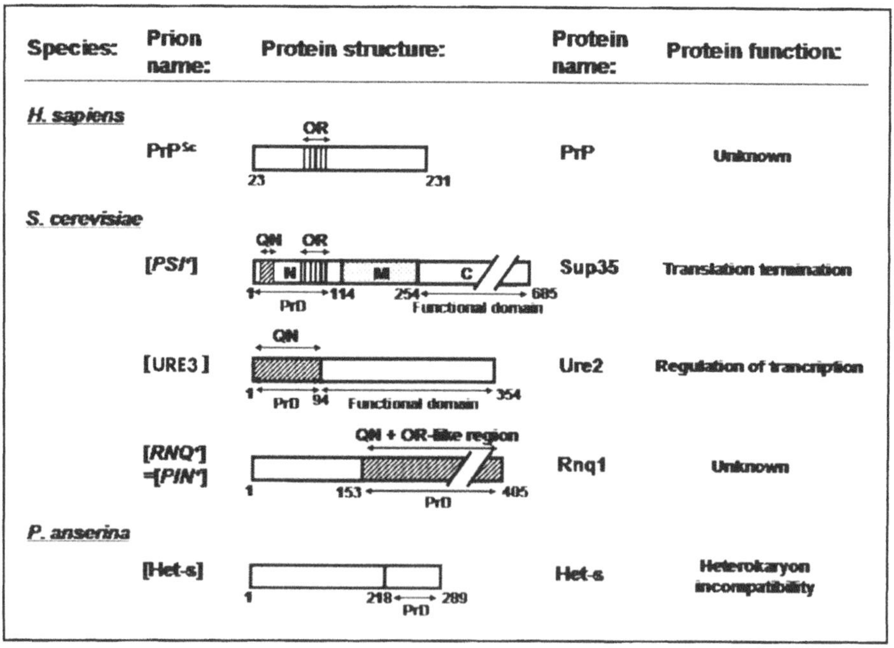

Figure 1. Structural organization of prion proteins. QN: the QN-rich stretch. OR: the region of oligopeptide repeats. PrD—prion domain. Numbers correspond to amino acid (aa) positions. Arrows indicate domain and subdomain boundaries. N, M and C—N-proximal, middle and C-proximal regions of Sup35, respectively. The N/M and M/C boundaries are arbitrarily assigned to the second (aa 124) and third (aa 254) methionine residues of the Sup35 protein. See text for details.

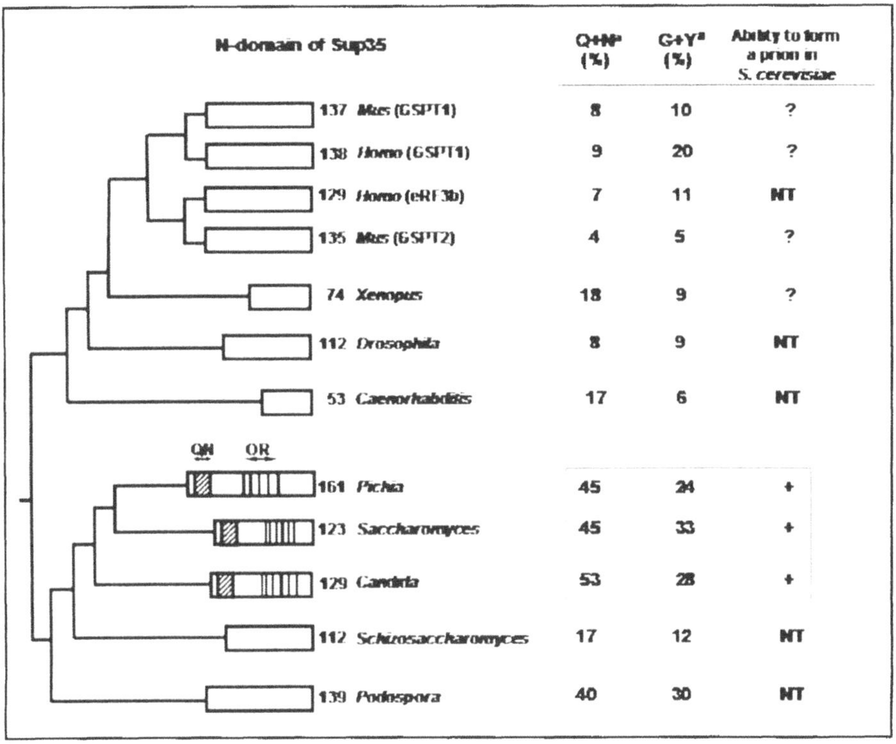

Figure 2. Evolutionary comparison of the N-terminal domains of Sup35 homologs. Sequences are from http://www.ncbi.nlm.nih.gov. Taxonomical relationships are from http://www.ncbi.nlm.nih.gov/Taxonomy. Scales do not correspond to evolutionary distances. QN and OR—see Figure 1. Numbers on the right correspond to the size of the N-terminal region (in aa) in each case. (Q + N) and (G + Y) content of N-terminal domain of Sup35 homologs was calculated using aa composition program at http://npsa-pbil.ibcp.fr. Sequence data were obtained from http://www.ncbi.nlm.nih.gov. ?—refers to the cases where search for prion activity in *S. cerevisiae* has been performed but have not yielded positive results (O. Zemlyanko, A. Petrova and G. Zhouravleva, unpublished; K. Gokhale and Y. Chernoff, unpublished). NT, not tested.

(10%, 15% and 20%, respectively). Instead of the QN and OR, mammalian eRF3 proteins contain poly-G and/or poly-S tracts. In mammals with two different eRF3-coding genes, all GSPT1 orthologs contain both poly-G and poly-S, while GSPT2 orthologs contain only poly-S. These homopeptide regions are usually coded almost exclusively by identical repeated trinucleotides, suggesting that they originate from trinucleotide expansions. Recent data confirm that the poly-G expansion can indeed occur in GSPT1 and is associated with susceptibility to gastric cancer.[67] Obviously eRF3 homologs of higher eukaryotes possess some unusual properties, although it remains to be seen whether these properties involve an ability to form amyloids.

At the current level of knowledge, it can not be ruled out that conservation of the Sup35N aa composition in budding yeast or unusual features of the aa composition of this region in other organisms are associated with its unknown function that is not directly related to prion formation. A variety of cellular proteins interact with Sup35N and/or Sup35M regions.[5] It is possible that Sup35N influences a function of the whole protein or targets it to a specific cell compartment. Indeed, the deletion of Sup35NM coding region leads to an alteration of the sexual cycle in *Podospora*,[68] implying that this region is not completely irrelevant to the cellular function of the protein.

Prion Role in Self/Nonself Recognition: Example of the [Het-s] Prion in Podospora

The first example of a prion having an adaptive biological function is [Het-s] of *Podospora* that controls vegetative incompatibility[69] (see also chapter 3). A cytoplasmic contact between the prion-containing and prion-free mycelia results in degeneration of the latter one. In this way, [Het-s] controls vegetative incompatibility, an adaptive trait in *Podospora*. Moreover, after meiotic division [Het-s] prion kills spores containing a *het-S* allele that is incapable of producing the prion state.[70] [Het-s] is abundant in natural *Podospora* populations. As adaptive function of [Het-s] is achieved via cytotoxic effect, [Het-s] combines features of both "prion pathology" and "adaptive prionization" models.

Role of [Het-s] in cytoplasmic incompatibility is related to one general characteristic feature of amyloids, that is, to a high level of sequence-specificity in amyloid propagation. While proteins of different sequences may possess amyloid properties, only molecules that contain the amyloid-forming domains of nearly identical sequences can join any given amyloid fiber. Recent data show that at least some amyloids are assembled together via parallel β-sheets, for which identity of aa sequences involved in β-sheet formation is extremely important.[11,71] In terms of their stringency, sequence identity requirements for amyloid formation are not dissimilar from the rules that govern complementarity of DNA strands. These requirements may explain so-called "species barrier" in prion transmission, preventing transmission of the prion state between the divergent prion domains (reviewed in ref. 65). Sequence-specificity makes prions a useful tool for the self/nonself recognition systems, as demonstrated by the example of cytoplasmic incompatibility in *Podospora*.

Stress Granules and Protection against Stresses

In higher eukaryotes, the stress such as heat shock is followed by formation of the nuclear and/or cytoplasmic stress granules (SG).[72] Cytoplasmic SGs contain transcripts associated with 40S ribosomal subunits (48S complexes), unable to initiate translation in stress conditions. SG assembly is mediated by the RNA-binding protein TIA-1,[73] which contains the C-terminal RNA recognition motif and Q-rich domain (Fig. 3A) similar to prion domains of yeast prion proteins. Deletion of Q-rich domain blocks SG formation after arsenite-induced stress in the mammalian cell culture, while substitution of TIA-1 "prion" domain for Sup35 prion domain (PrD) restores SG production. However in contrast to prion formation, TIA-1 aggregation and SG assembly are reversible after return to normal conditions[72] (Fig. 3B). Therefore, SGs provide an example of labile and economical posttrancriptional regulatory and protective mechanism contributing to the cellular function in stress conditions and based on prion-like properties.

There are several other examples of the protective mechanisms based on amyloid properties. Embryos of the fish *Austrofundulus limnaeus* are surrounded by an egg envelope composed of two proteins that together form a structure similar to amyloid fibrils.[74] Another fish protein, type I antifreeze protein that is normally α-helical, is converted into an amyloid upon freezing, that may possibly play a protective role by inhibiting ice formation.[75] As aggregation of the yeast prion proteins is increased in the stationary or nondividing cells,[54,76,77] one attractive speculation is that reversible PrD-mediated aggregation is used to protect some important proteins (e.g., Sup35) during unfavorable conditions.

Other Biological Roles of Amyloid-Like Structures

Ability of prions to fix and "memorize" protein conformational changes make them ideal candidates for the role of memory molecules. Indeed, it has been hypothesized that a prion-like domain of the neuron-specific isoform of cytoplasmic polyadenylation element binding protein (CPEB) is connected to long-term memory in the shellfish *Aplysia*.[78]

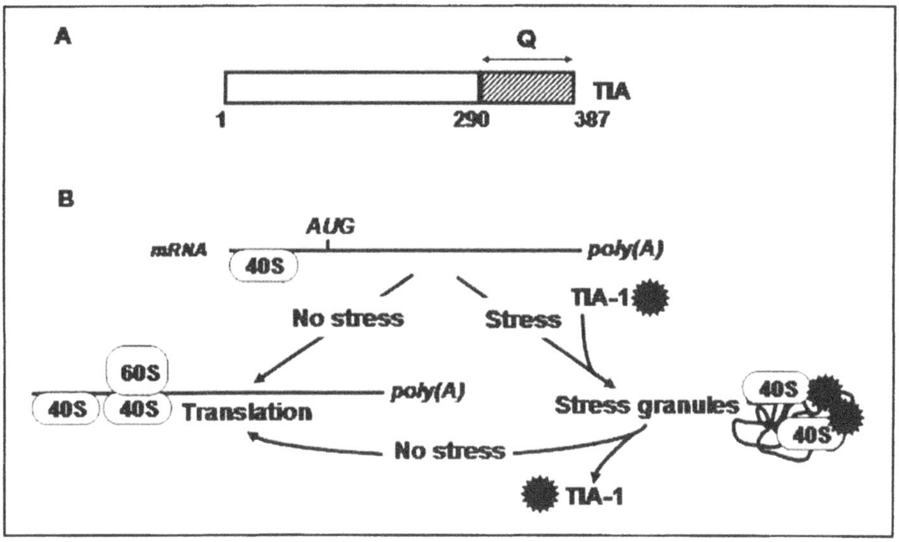

Figure 3. Formation of the stress granules. A) Schematic structure of TIA protein. Q: the Q-rich stretch. Other designations are as in Figure 1B Model showing formation of stress granules. Ribosome subunits are shown as ovals and TIA as black asterisk. See text for more details.

There is a number of other polymerized proteins that exhibit similarities to amyloid fibers, for example the spider silk protein, spidroin,[79] whose adaptive role in spiders is evident.[80] It has recently been shown that one of the mammalian proteins involved in melanin production adopts an amyloid structure, so that amyloid polymers likely serve as a scaffold for melanin polymerization[81] (Fig. 4). This is a first clear evidence for the positive biological role of amyloids in mammals, although it is not known whether this specific kind of amyloid possesses prion properties.

There also are examples of a beneficial role of amyloid-like aggregates in bacteria, such as facilitation of biofilm formation in *E. coli* by the extracellular self-assembly of the major curli protein, CsgA, containing PrP-like oligopeptide repeats,[82] into typical amyloid fibrils.[83] Amyloid-forming proteins of *Streptomyces coelicolor*, called chaplins, are essential for aerial growth.[84] Moreover, it has been hypothesized[85] that amyloid-like formations played an important role in the emergence of the primordial membranes and other structures at the early steps of the biological compartmentalization (reviewed in ref. 7)

Possible Evolutionary Consequences of Sup35 Prionization

Numerous attempts to identify an adaptive function of the prion state were made in case of Sup35 (eRF3), which is a translation termination factor. Formation of [*PSI*$^+$] prion decreases supply of functional Sup35, leading to efficient readthrough of the nonsense-mutations within ORFs. It remains unclear to which extent termination at the normal terminators, usually protected by nucleotide context,[86] is affected by [*PSI*$^+$]. In some genotypic backgrounds, presence of [*PSI*$^+$] induces heat shock response[87] and increases resistance of yeast cells to some stresses.[88] Although "protective" in the artificially generated laboratory situations, such abnormalities in Hsp levels would not likely be adaptive in the long run in nature.

Systematic comparison of a variety of phenotypes (such as resistance to certain toxicants, etc.) between several isogenic pairs of [*PSI*$^+$] and [*psi*$^-$] strains has shown that the presence of [*PSI*$^+$] was beneficial in some conditions for certain genotypes.[23] However, ancestors of these laboratory strains went through multiple rounds of mutagenesis and could therefore contain

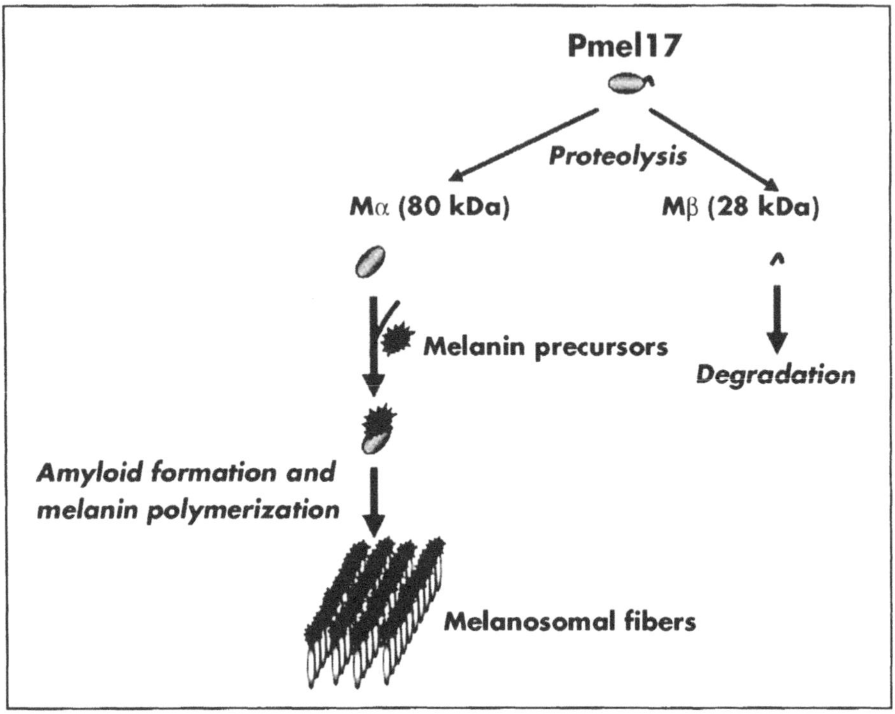

Figure 4. Role of amyloid in melanin polymerization. Glycoprotein Pmel17, that is a critical component of melanosome biogenesis, gives rise to two fragments, Mα and Mβ. Self-assembly of Mα leads to amyloid formation. Amyloid provides a scaffold for melanin polymerization.

unidentified nonsense-alleles. While suppression of such alleles could be beneficial for these specific strains in the laboratory, the question remains whether or not this is directly applicable to natural conditions.

It was proposed[23] that the presence of [PSI^+] could increase the "evolvability" of the yeast population and facilitate adaptation to environmental changes by generating new protein products from ORFs containing nonsense-mutations, weak terminators or frameshifting-prone sequences. Such a mechanism could in principle be applied to activation of the silent pseudogenes.[89,90] As an extension of the modular principle in molecular evolution,[91] one could suggest that new genes can be created through recombination of inactivated (pseudogene) copies, which often have no introns and are "locked" by nonsense and frameshift mutations. As pseudogenes are not functional, they can easily accumulate new mutations potentially generating new functions.[92] Sporadic activation of pseudogenes through nonsense or frameshift suppression allows natural selection to choose combinations of mutations having beneficial effects. Analysis of whole genomes has revealed a number of cases which can serve as examples of possible pseudogene resurrection.[93]

If [PSI^+] decreases termination efficiency and therefore allow pseudogene expression, such read-through events may take place at a frequency of at least one per every million years, as suggested by the quantitative model.[94] However, mutations in the genes coding for the components of translation machinery may have the same effect.[5] It is therefore not clear whether the proposed mechanism is specific to a prion. Although mutated translational components are likely to turn detrimental in natural environments, so is [PSI^+], judged from analysis of the natural yeast isolates.[22,43,44]

One potential advantage of [*PSI*⁺], not shared by most of the abovementioned gene mutants, could be that it is a dominant omnipotent suppressor affecting both termination and frameshifting. Another possibility that would give [*PSI*⁺] a preference at the population level over other mechanisms causing nonsense readthrough could be an easier transition between [*psi*⁻] and [*PSI*⁺] states. However, frequencies of spontaneous acquisition and loss of the typical "strong" [*PSI*⁺] variants are quite low, making them unlikely candidates for such a role. It is therefore possible that increased adaptability could be associated not with the "conventional" stable prion variants used in most laboratory experiments, but with the variants maintained only in certain conditions and eliminated after conditions are changed. Proof of the existence of such conditionally stable [*PSI*⁺] variants has been provided recently by identification of the [*PSI*⁺] isolate that can be maintained only at high levels of the chaperone Hsp104.[95] It still remains to be shown which (if any) conditions in nature could favor maintenance of such transient variants of [*PSI*⁺].

Conclusion: Prions, Protein Mutants and Posttranslational Feedback Regulation

While strong experimental data support "prion pathology" model, evidence in favor of the "adaptive prionization" model is of rather circumstantional nature. Most examples of the proven biologically positive effects of amyloid-like formations (melanin biosynthesis, stress granules, etc.) are so far dealing with the nonprion aggregates. The only prion which is clearly documented to play a biologically positive role in natural conditions, [Het-s] of *Podospora*, ironically does so by killing a nonprion partner.

However, one should remember that the majority of the known prions were identified by chance, due to extreme phenotypic effects caused by the corresponding proteins in the prion form, such as fatal transmissible disease in case of mammalian PrP or translation termination defect in case of yeast Sup35. It is possible that we are so far dealing only with a very top of the iceberg, and a large number of prion-like phenomena are still waiting for their discoverers. If prions are to be considered as "mutants" occurring at the protein level,[6] one should not expect that randomly chosen mutations would frequently turn beneficial for the organism. Rather, the majority of them would be expected to have either deleterious effect or no effect, as in case of DNA mutations. However, it does not exclude a possibility of some beneficial changes occurring by this mechanism that could be identified in the future.

Another possible dimension of this story is that beneficial effects could be associated with the transient prion variants, as hypothesized above in case of [*PSI*⁺], while the stably propagating and usually toxic prions might represent by-products of these processes. Normal cellular functions of Sup35 and Ure2 are decreased in the prion state, suggesting that transient formation of the prion-like multimers may serve as a mechanism of feedback regulation. This notion is supported by the existence of the shortened form of Ure2, generated by alternative translational initiation and lacking the prion domain.[96] Likewise, existence of the shortened transcript of the *SUP35* gene in certain conditions has been reported[97] but never studied carefully. Many proteins involved in DNA replication, repair and transcription contain PrD-like QN-rich domains.[98] In case of the yeast transcriptional repressor Gal11, existence of two alternative transcripts has been demonstrated, of which the shorter one is missing two QN-rich domains and codes for the protein that manifests itself as a transcriptional activator rather than repressor.[99] These data suggest that prion-like mechanisms of feedback regulation could be widespread, and this may explain evolutionary conservation of prion properties.

One should note that transient prion variants maintained only in certain conditions are hard to distinguish from both feedback regulatory circuits and so-called "heritable" modifications persisting for a few generations. Therefore, role of the transient prion variants in adaptive evolution, as hypothesized above, would be in agreement with the more general hypothesis of V. Kirpichnikov[100] regarding the role of modifications in evolution. Moreover, prion model may provide a tool for even more direct relationship between phenotypic and "genotypic" (in traditional sense) inheritance. As prion state of a protein may influence probability of prionization of another protein[17-19] (see also chapter 4), this opens a possibility for concerted modification (prionization) of several proteins at once. Such a prionization network, in turn, may potentially influence a DNA metabolism and rate of "classic" mutations, in case if some of the prionized proteins are involved in DNA replication/repair. This provides a mechanism for the possible effects of the heritable protein variations on the DNA material.

Acknowledgements

We thank R.B. Wickner and G.P. Newnam for critical reading of the manuscript and helpful suggestions. This work was supported by grants ST-012 from CRDF, RAS Presidium Program "Biosphere origin and evolution" and (Lot 2006-12.2/001) from Federal agency of science and innovations (to S.G. I.-V. and G.A.Z.), and by grant R01GM58763 from NIH (to Y.O.C.).

References

1. Crick FH. On protein synthesis. Symp Soc Exp Biol 1958; 12:138-163.
2. Koltsov NK. Inherent molecules. Cell Organisation. Moscow-Leningrad: State Publishing House of Biol and Med Lit, 1936:586-620.
3. Prusiner SB, Scott MR, Dearmond SJ et al. Prion protein biology. Cell 1998; 93:337-348.
4. Wickner RB, Edskes HK, Ross ED et al. Prion genetics: New rules for a new kind of gene. Annu Rev Genet 2004; 38:681-707.
5. Inge-Vechtomov S, Zhouravleva G, Philippe M. Eukaryotic release factors (eRFs) history. Biol Cell 2003; 95:195-209.
6. Chernoff YO. Mutation processes at the protein level: Is Lamarck back? Mutat Res 2001; 488:39-64.
7. Chernoff YO. Amyloidogenic domains, prions and structural inheritance: Rudiments of early life or recent acquisition? Curr Opin Chem Biol 2004; 8:665-671.
8. Diaz-Avalos R, Long C, Fontano E et al. Cross-beta order and diversity in nanocrystals of an amyloid-forming peptide. J Mol Biol 2003; 330:1165-1175.
9. Krishnan R, Lindquist SL. Structural insights into a yeast prion illuminate nucleation and strain diversity. Nature 2005; 435:765-772.
10. Glover JR, Kowal AS, Schirmer EC et al. Self-seeded fibers formed by Sup35, the protein determinant of [PSI⁺], a heritable prion-like factor of S. cerevisiae. Cell 1997; 89:811-819.
11. Nelson R, Sawaya MR, Balbirnie M et al. Structure of the cross-beta spine of amyloid-like fibrils. Nature 2005; 435:773-778.
12. Tanaka M, Chien P, Yonekura K et al. Mechanism of cross-species prion transmission: An infectious conformation compatible with two highly divergent yeast prion proteins. Cell 2005; 121:49-62.
13. King CY, Diaz-Avalos R. Protein-only transmission of three yeast prion strains. Nature 2004; 428:319-323.
14. Tanaka M, Chien P, Naber N et al. Conformational variations in an infectious protein determine prion strain differences. Nature 2004; 428:323-328.
15. Jeffrey M, Gonzalez L. Pathology and pathogenesis of bovine spongiform encephalopathy and scrapie. Curr Top Microbiol Immunol 2004; 284:65-97.
16. Aguzzi A, Polymenidou M. Mammalian prion biology: One century of evolving concepts. Cell 2004; 116:313-327.
17. Derkatch IL, Bradley ME, Zhou P et al. Genetic and environmental factors affecting the de novo appearance of the [PSI⁺] prion in Saccharomyces cerevisiae. Genetics 1997; 147:507-519.
18. Derkatch IL, Bradley ME, Hong JY et al. Prions affect the appearance of other prions: The story of [PIN⁺]. Cell 2001; 106:171-182.
19. Osherovich LZ, Weissman JS. Multiple Gln/Asn-rich prion domains confer susceptibility to induction of the yeast [PSI⁺] prion. Cell 2001; 106:183-194.
20. Dobson CM. Protein folding and misfolding. Nature 2003; 426:884-890.
21. Stefani M, Dobson CM. Protein aggregation and aggregate toxicity: New insights into protein folding, misfolding diseases and biological evolution. J Mol Med 2003; 81:678-699.
22. Nakayashiki T, Kurtzman CP, Edskes HK et al. Yeast prions [URE3] and [PSI⁺] are diseases. Proc Natl Acad Sci USA 2005; 102:10575-10580.
23. True HL, Lindquist SL. A yeast prion provides a mechanism for genetic variation and phenotypic diversity. Nature 2000; 407:477-483.

24. Horwich AL, Weissman JS. Deadly conformations: Protein misfolding in prion disease. Cell 1997; 89:499-510.
25. Dobson CM. Protein misfolding, evolution and disease. Trends Biochem Sci 1999; 24:329-332.
26. Pepys MB, Hawkins PN, Booth DR et al. Human lysozyme gene mutations cause hereditary systemic amyloidosis. Nature 1993; 362:553-557.
27. Faux NG, Bottomley SP, Lesk AM et al. Functional insights from the distribution and role of homopeptide repeat-containing proteins. Genome Res 2005; 15:537-551.
28. Orr HT. Beyond the Qs in the polyglutamine diseases. Genes Dev 2001; 15:925-932.
29. Dorsman JC, Pepers B, Langenberg D et al. Strong aggregation and increased toxicity of polyleucine over polyglutamine stretches in mammalian cells. Hum Mol Genet 2002; 11:1487-1496.
30. Fandrich M, Dobson CM. The behaviour of polyamino acids reveals an inverse side chain effect in amyloid structure formation. EMBO J 2002; 21:5682-5690.
31. Oma Y, Kino Y, Sasagawa N et al. Intracellular localization of homopolymeric amino acid-containing proteins expressed in mammalian cells. J Biol Chem 2004; 279:21217-21222.
32. Kihm AJ, Kong Y, Hong W et al. An abundant erythroid protein that stabilizes free alpha-haemoglobin. Nature 2002; 417:758-763.
33. Bueler H, Aguzzi A, Sailer A et al. Mice devoid of PrP are resistant to scrapie. Cell 1993; 73:1339-1347.
34. Gidalevitz T, Ben Zvi A, Ho KH et al. Progressive disruption of cellular protein folding in models of polyglutamine diseases. Science 2006; 311:1471-1474.
35. Scherzinger E, Sittler A, Schweiger K et al. Self-assembly of polyglutamine-containing huntingtin fragments into amyloid-like fibrils: Implications for Huntington's disease pathology. Proc Natl Acad Sci USA 1999; 96:4604-4609.
36. Steffan JS, Kazantsev A, Spasic-Boskovic O et al. The Huntington's disease protein interacts with p53 and CREB-binding protein and represses transcription. Proc Natl Acad Sci USA 2000; 97:6763-6768.
37. Steffan JS, Bodai L, Pallos J et al. Histone deacetylase inhibitors arrest polyglutamine-dependent neurodegeneration in Drosophila. Nature 2001; 413:739-743.
38. Satyal SH, Schmidt E, Kitagawa K et al. Polyglutamine aggregates alter protein folding homeostasis in Caenorhabditis elegans. Proc Natl Acad Sci USA 2000; 97:5750-5755.
39. Leblanc AC. The role of apoptotic pathways in Alzheimer's disease neurodegeneration and cell death. Curr Alzheimer Res 2005; 2:389-402.
40. Liberski PP, Sikorska B, Bratosiewicz-Wasik J et al. Neuronal cell death in transmissible spongiform encephalopathies (prion diseases) revisited: From apoptosis to autophagy. Int J Biochem Cell Biol 2004; 36:2473-2490.
41. Rego AC, Oliveira CR. Mitochondrial dysfunction and reactive oxygen species in excitotoxicity and apoptosis: Implications for the pathogenesis of neurodegenerative diseases. Neurochem Res 2003; 28:1563-1574.
42. Roucou X, Leblanc AC. Cellular prion protein neuroprotective function: Implications in prion diseases. J Mol Med 2005; 83:3-11.
43. Chernoff YO, Galkin AP, Lewitin E et al. Evolutionary conservation of prion-forming abilities of the yeast Sup35 protein. Mol Microbiol 2000; 35:865-876.
44. Resende CG, Outeiro TF, Sands L et al. Prion protein gene polymorphisms in Saccharomyces cerevisiae. Mol Microbiol 2003; 49:1005-1017.
45. Cox BS, Tuite MF, McLaughlin CS. The psi factor of yeast: A problem in inheritance. Yeast 1988; 4:159-178.
46. Fabrizio P, Battistella L, Vardavas R et al. Superoxide is a mediator of an altruistic aging program in Saccharomyces cerevisiae. J Cell Biol 2004; 166:1055-1067.
47. Herker E, Jungwirth H, Lehmann KA et al. Chronological aging leads to apoptosis in yeast. J Cell Biol 2004; 164:501-507.
48. Madeo F, Herker E, Maldener C et al. A caspase-related protease regulates apoptosis in yeast. Mol Cell 2002; 9:911-917.
49. Bradley ME, Edskes HK, Hong JY et al. Interactions among prions and prion "strains" in yeast. Proc Natl Acad Sci USA 2002; 99:16392-16399.
50. Chernoff YO, Inge-Vechtomov SG, Derkach IL et al. Dosage-dependent translational suppression in yeast Saccharomyces cerevisiae. Yeast 1992; 8:489-499.
51. Dagkesamanskaya AR, Ter Avanesyan MD. Interaction of the yeast omnipotent suppressors SUP1 (SUP45) and SUP2 (SUP35) with non-Mendelian factors. Genetics 1991; 128:513-520.
52. Derkatch IL, Chernoff YO, Kushnirov VV et al. Genesis and variability of [PSI] prion factors in Saccharomyces cerevisiae. Genetics 1996; 144:1375-1386.

53. Ganusova EE, Ozolins LN, Bhagat S et al. Modulation of prion formation, aggregation, and toxicity by the actin cytoskeleton in yeast. Mol Cell Biol 2006; 26:617-629.
54. Zhou P, Derkatch IL, Liebman SW. The relationship between visible intracellular aggregates that appear after overexpression of Sup35 and the yeast prion-like elements [PSI+] and [PIN+]. Mol Microbiol 2001; 39:37-46.
55. Meriin AB, Zhang X, He X et al. Huntington toxicity in yeast model depends on polyglutamine aggregation mediated by a prion-like protein Rnq1. J Cell Biol 2002; 157:997-1004.
56. Gokhale KC, Newnam GP, Sherman MY et al. Modulation of prion-dependent polyglutamine aggregation and toxicity by chaperone proteins in the yeast model. J Biol Chem 2005; 280:22809-22818.
57. Meriin AB, Zhang X, Miliaras NB et al. Aggregation of expanded polyglutamine domain in yeast leads to defects in endocytosis. Mol Cell Biol 2003; 23:7554-7565.
58. Bailleul PA, Newnam GP, Steenbergen JN et al. Genetic study of interactions between the cytoskeletal assembly protein Sla1 and prion-forming domain of the release factor Sup35 (eRF3) in Saccharomyces cerevisiae. Genetics 1999; 153:81-94.
59. Flower TR, Chesnokova LS, Froelich CA et al. Heat shock prevents alpha-synuclein-induced apoptosis in a yeast model of Parkinson's disease. J Mol Biol 2005; 351:1081-1100.
60. Kushnirov VV, Kochneva-Pervukhova NV, Chechenova MB et al. Prion properties of the Sup35 protein of yeast Pichia methanolica. EMBO J 2000; 19:324-331.
61. Nakayashiki T, Ebihara K, Bannai H et al. Yeast [PSI+] "prions" that are crosstransmissible and susceptible beyond a species barrier through a quasi-prion state. Mol Cell 2001; 7:1121-1130.
62. Santoso A, Chien P, Osherovich LZ et al. Molecular basis of a yeast prion species barrier. Cell 2000; 100:277-288.
63. Zadorskii SP, Sopova I, Inge-Vechtomov SG. Prionization of the Pichia methanolica SUP35 gene product in the yeast Saccharomyces cerevisiae. Genetika 2000; 36:1322-1329.
64. Jensen MA, True HL, Chernoff YO et al. Molecular population genetics and evolution of a prion-like protein in Saccharomyces cerevisiae. Genetics 2001; 159:527-535.
65. Zhouravleva G, Alenin V, Inge-Vechtomov S et al. To stick or not to stick: Prion domains from yeast to mammals. In: Pandalai SG, ed. Recent Res Devel Mol Cell Biol 2002; 3(Pt 1):185-218.
66. Ross ED, Edskes HK, Terry MJ et al. Primary sequence independence for prion formation. Proc Natl Acad Sci USA 2005; 102:12825-12830.
67. Brito M, Malta-Vacas J, Carmona B et al. Polyglycine expansions in eRF3/GSPT1 are associated with gastric cancer susceptibility. Carcinogenesis 2005; 26:2046-2049.
68. Gagny B, Silar P. Identification of the genes encoding the cytosolic translation release factors from Podospora anserina and analysis of their role during the life cycle. Genetics 1998; 149:1763-1775.
69. Coustou V, Deleu C, Saupe S et al. The protein product of the het-s heterokaryon incompatibility gene of the fungus Podospora anserina behaves as a prion analog. Proc Natl Acad Sci USA 1997; 94:9773-9778.
70. Dalstra HJ, Swart K, Debets AJ et al. Sexual transmission of the [Het-s] prion leads to meiotic drive in Podospora anserina. Proc Natl Acad Sci USA 2003; 100:6616-6621.
71. Ross ED, Minton A, Wickner RB. Prion domains: Sequences, structures and interactions. Nat Cell Biol 2005; 7:1039-1044.
72. Anderson P, Kedersha N. Stressful initiations. J Cell Sci 2002; 115:3227-3234.
73. Gilks N, Kedersha N, Ayodele M et al. Stress granule assembly is mediated by prion-like aggregation of TIA-1. Mol Biol Cell 2004; 15:5383-5398.
74. Podrabsky JE, Carpenter JF, Hand SC. Survival of water stress in annual fish embryos: Dehydration avoidance and egg envelope amyloid fibers. Am J Physiol Regul Integr Comp Physiol 2001; 280:R123-R131.
75. Graether SP, Slupsky CM, Sykes BD. Freezing of a fish antifreeze protein results in amyloid fibril formation. Biophys J 2003; 84:552-557.
76. Paushkin SV, Kushnirov VV, Smirnov VN et al. Propagation of the yeast prion-like [psi+] determinant is mediated by oligomerization of the SUP35-encoded polypeptide chain release factor. EMBO J 1996; 15:3127-3134.
77. Bailleul-Winslett PA, Newnam GP, Wegrzyn RD et al. An antiprion effect of the anticytoskeletal drug latrunculin A in yeast. Gene Expr 2000; 9:145-156.
78. Si K, Lindquist S, Kandel ER. A neuronal isoform of the aplysia CPEB has prion-like properties. Cell 2003; 115:879-891.
79. Kenney JM, Knight D, Wise MJ et al. Amyloidogenic nature of spider silk. Eur J Biochem 2002; 269:4159-4163.
80. Craig CL. Spider webs and silks: Tracing evolution from molecules to genes to phenotypes. New York: Oxford University Press, 2003.

81. Fowler DM, Koulov AV, Alory-Jost C et al. Functional amyloid formation within mammalian tissue. PLoS Biol 2005; 4:e6.
82. Cherny I, Rockah L, Levy-Nissenbaum O et al. The formation of Escherichia coli curli amyloid fibrils is mediated by prion-like peptide repeats. J Mol Biol 2005; 352:245-252.
83. Chapman MR, Robinson LS, Pinkner JS et al. Role of Escherichia coli curli operons in directing amyloid fiber formation. Science 2002; 295:851-855.
84. Claessen D, Rink R, de Jong W et al. A novel class of secreted hydrophobic proteins is involved in aerial hyphae formation in Streptomyces coelicolor by forming amyloid-like fibrils. Genes Dev 2003; 17:1714-1726.
85. Zhang S, Holmes T, Lockshin C et al. Spontaneous assembly of a self-complementary oligopeptide to form a stable macroscopic membrane. Proc Natl Acad Sci USA 1993; 90:3334-3338.
86. Bonetti B, Fu L, Moon J et al. The efficiency of translation termination is determined by a synergistic interplay between upstream and downstream sequences in Saccharomyces cerevisiae. J Mol Biol 1995; 251:334-345.
87. Jung G, Jones G, Wegrzyn RD et al. A role for cytosolic hsp70 in yeast [PSI$^+$] prion propagation and [PSI$^+$] as a cellular stress. Genetics 2000; 156:559-570.
88. Eaglestone SS, Cox BS, Tuite MF. Translation termination efficiency can be regulated in Saccharomyces cerevisiae by environmental stress through a prion-mediated mechanism. EMBO J 1999; 18:1974-1981.
89. Inge-Vechtomov SG. A possible role of genetic translation ambiguity in evolution. Mol Biol (Mosk) 2002; 36:268-276.
90. Shorter J, Lindquist S. Prions as adaptive conduits of memory and inheritance. Nat Rev Genet 2005; 6:435-450.
91. Gilbert W. Why genes in pieces? Nature 1978; 271:501.
92. Koch AL. Enzyme evolution. I. The importance of untranslatable intermediates. Genetics 1972; 72:297-316.
93. Harrison P, Kumar A, Lan N et al. A small reservoir of disabled ORFs in the yeast genome and its implications for the dynamics of proteome evolution. J Mol Biol 2002; 316:409-419.
94. Masel J, Bergman A. The evolution of the evolvability properties of the yeast prion [PSI$^+$]. Evolution Int J Org Evolution 2003; 57:1498-1512.
95. Borchsenius AS, Muller S, Newnam GP et al. Prion variant maintained only at high levels of the Hsp104 disaggregase. Curr Genet 2006; 49:21-29.
96. Komar AA, Lesnik T, Cullin C et al. Internal initiation drives the synthesis of Ure2 protein lacking the prion domain and affects [URE3] propagation in yeast cells. EMBO J 2003; 22:1199-1209.
97. Surguchov AP, Telkov MV, Smirnov VN. Absence of structural homology between sup1 and sup2 genes of yeast Saccharomyces cerevisiae and identification of their transcripts. FEBS Lett 1986; 206:147-150.
98. Michelitsch MD, Weissman JS. A census of glutamine/asparagine-rich regions: Implications for their conserved function and the prediction of novel prions. Proc Natl Acad Sci USA 2000; 97:11910-11915.
99. Ono B, Futase T, Honda W et al. The Saccharomyces cerevisiae ESU1 gene, which is responsible for enhancement of termination suppression, corresponds to the 3'-terminal half of GAL11. Yeast 2005; 22:895-906.
100. Kirpichnikov VS. Role of noninherent variability in the process of natural selection. Biological Journal 1935; 4:775-801.

CHAPTER 9

Preformed Cell Structure and Cell Heredity

Janine Beisson*

Abstract

This chapter will first recall the phenomena of "cortical inheritance" observed and genetically demonstrated in *Paramecium* 40 years ago, and later in other ciliates (Tetrahymena, Oxytricha, Paraurostyla), and will analyze the deduced concept of "cytotaxis" or "structural memory". The significance of these phenomena, all related (but not strictly restricted) to the properties of ciliary basal bodies and their mode of duplication, will be interpreted in the light of present knowledge on the mechanism and control of basal body/ centriole duplication. Then other phenomena described in a variety of organisms will be analyzed or mentioned which show the relevance of the concept of cytotaxis or structural memory to other cellular processes, mainly (1) cytoskeleton assembly and organization with examples on ciliates, trypanosome, mammalian cells and plants, and (2) transmission of polarities with examples on yeast, trypanosome and metazoa. Finally, I will discuss some aspects of this particular type of non-DNA inheritance: (1) why so few documented examples if structural memory is a basic parameter in cell heredity, and (2) how are these phenomena (which all rely on protein/protein interactions, and imply a formatting role of preexisting proteinic complexes on neo-formed proteins and their assembly) related to prions?

Introduction

Our understanding of biological processes is constrained within a circular relationship resembling the "hen-egg dilemma" which might be called the "DNA-cell dilemma": if all the information necessary to make a cell (or a tyrannosaurus, as the fiction goes) is stored in DNA sequences, why does a cell only arise from a preexisting cell? The uninterrupted cell continuity since LUCA suggests that cell heredity might require more than DNA. Aside from membranes which, like DNA, cannot form de novo, what does a cell transmit to its daughters that allows them to recapitulate the exact morphology of their mother, despite the profound remodeling that accompanies division?

Such a question may come to the mind of any biologist watching his/her favourite cell dividing; however, some cell types are undoubtedly more suggestive of the existence of a cellular memory than others, and it may not be fortuitous that a ciliate, *Paramecium*, was the first organism to inspire a genetic approach of the problem.[1] Ciliates are unicellular organisms characterized by the elaborate asymmetrical organization of their surface (cortex), with the equivalent of multiple different organs, arranged in a specific body plan, so that each division involves complex morphogenetic movements akin to developmental processes. The high fidelity of reproduction of every minute detail of this complex pattern raised two types of questions: Is cell organization genetically determined in all details? Can morphogenetic processes rely only on the specificity of proteins and self-assembly mechanisms or do they also involve the persistance of structural or biochemical landmarks to facilitate or template assembly?

*Janine Beisson—Centre de Génétique Moléculaire, Centre National de la Recherche Scientifique 91190 Gif-sur-Yvette, France. Email: beisson@cgm.cnrs-gif.fr

Protein-Based Inheritance, edited by Yury O. Chernoff. ©2007 Landes Bioscience.

In the early sixties, these questions were addressed by T.M. Sonneborn, the American biologist who had developped *Paramecium* as a model genetic system; he studied different spontaneous or experimental variations in *Paramecium* cortical organization and demonstrated their hereditary maintenance by cellular continuity, without any change in the DNA-encoded information. This cortical inheritance led to the concept of "cytotaxis", or "structural inheritance" at about the same time as the non-DNA based nature of the scrapie agent first began to be suspected.[2]

This chapter will analyse the experimental bases of this concept, evoke its multifarious interventions in cell heredity and try to discuss its general significance and biological importance as a manifestation of non-DNA-based inheritance.

Paramecium: A Model for a New Concept

Figure 1 depicts the organization of the cortex of *P.tetraurelia* featuring over 4000 ciliary basal bodies. A ciliate can best be thought of as a metazoon embryo which missed cellularization,

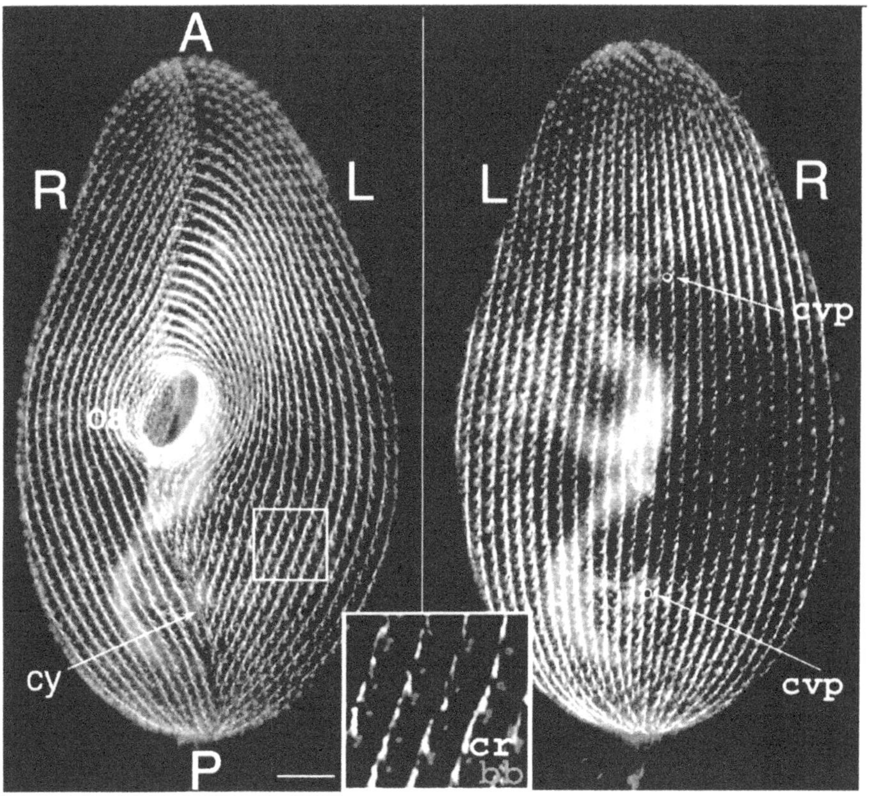

Figure 1. Cortical organization of *Paramecium*. The figure shows the ventral (left) and dorsal (right) sides of a cell immunolabeled by an anti-tubulin antibody which reveals basal bodies as discrete dots (bb) and an antibody directed against the ciliary rootlets (cr) which form a thin bundle emanating from each basal body, as shown in the enlagement. A-P marks the antero-posterior cell axis. The ventral side is marked by a line of contrast in the global arrangement of basal body rows, the oral meridian, which defines the right(R) and the left (L) of the cell. The oral apparatus (oa)—a ciliated funnel at the bottom of which phagocytosis takes place—and the cytoproct (cy) open on the ventral side, while the pores of the contractile vacuole systems (cvp) localize on the dorsal surface. Bar: 10 µm. Images: courtesy of F. Ruiz.

but has nevetheless partitioned its surface into repeated units, the cortical units, each with one or two basal bodies flanked by cytoskeletal appendages; the whole forms a mosaïc of territories endowed with distinct morphogenetic capacities.[3] In addition, diverse organelles fulfill specialized functions, e.g., a ciliated oral apparatus (oa) for phagocytosis, a contractile cytoproct (cy) for excretion of food vacuole residues, a complex vesicular and microtubular system for osmoregulation opening at contractile vacuole pores (cvp). Most importantly, the cell displays striking asymmetries. Cortical units align in rows parallel to the antero-posterior axis of the cell; on the ventral surface, the oral apparatus and cytoproct localize along an oral meridian which defines the axis of right-left asymmetry, while the dorsal side is marked by the contractile vacuole pores. The global right-left asymmetry and antero-posterior polarity of the cell are expressed at the cortical unit level by the direction of the cytoskeletal appendages nucleated around each basal body such as the conspicuous ciliary rootlet (Fig. 1).

During division (Fig. 2), the organelles are duplicated and a wave of basal body duplication pervades the cell. Like centrioles in centrosomes, basal bodies duplicate by a conservative process with each neo-formed organelle arising close and at right angles to the mother (Fig. 3).[4] For ciliary or flagellar basal bodies, the process obeys a strict polarity: the new basal body inserts in

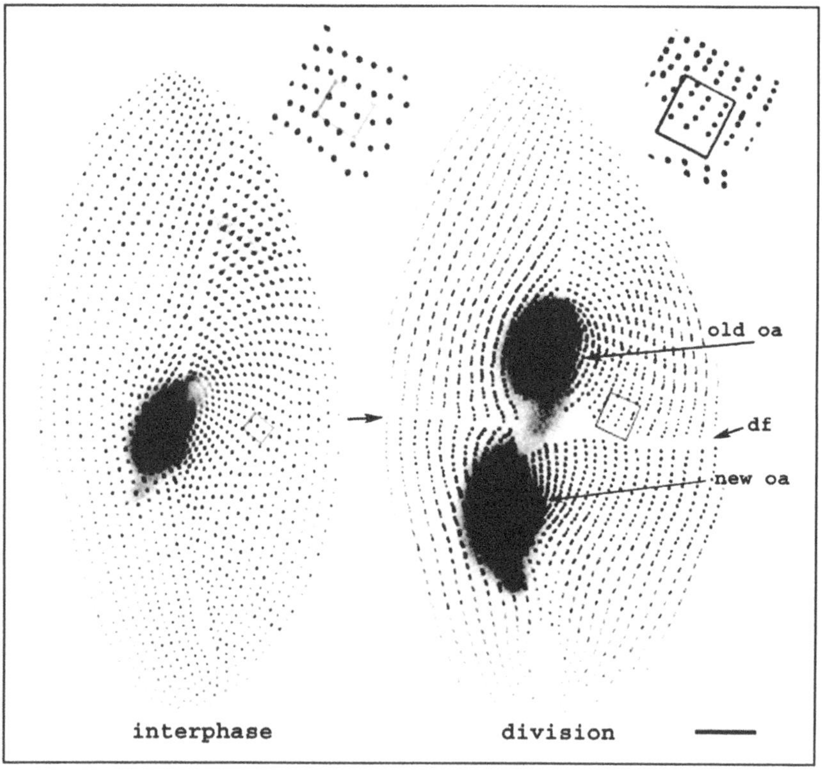

Figure 2. Morphogenetic processes during division. The figure shows interphase and dividing living cells expressing GFP-PtCen2a, a centrin specific to basal bodies.[29] Smaller dots correspond to a single basal body per cortical unit, larger ones to two basal bodies per unit. In the dividing cell, where basal body duplication proceeds, the division furrow (df) delimits the two presumptive daughter cells. The old oral apparatus is conserved in the anterior daughter cell, while a new one has developped in the posterior one. The two insets pinpoint three units with 2 bbs in the interphase cell and their progeny in the dividing cell. Images are shown with reverse contrast. Bar: 10 μm. Images: courtesy of F. Ruiz.

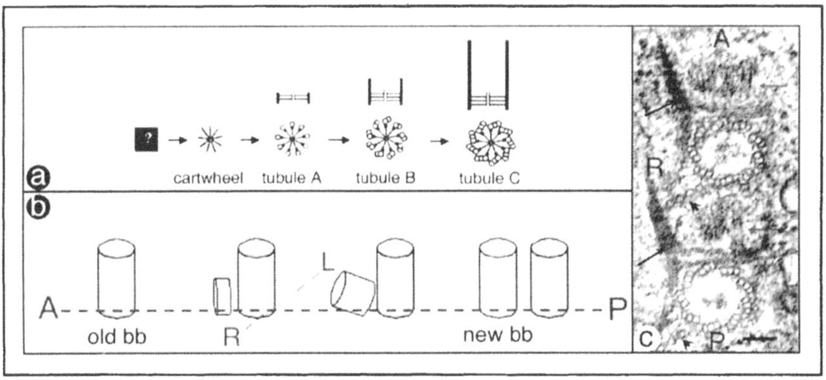

Figure 3. Basal body duplication. a) The assembly line. The recorded steps (here separated for clarity but which most probably overlap) correspond to the present state of knowledge: a black box (?) still shields the very initial steps; the 9-branched star symbolizes the required prepattern for the characteristic ninefold symmetry of the microtubule cylinder which elongates along with the sequential addition of tubules A, B, C. b) Polarity of basal body duplication. Each new basal body (new bb), assembled as described in a, develops at right angles and anterior to its mother (old bb), then tilts up to become inserted in the cortex, along the same basal body row as the mother organelle. c) Basal body duplication in situ. This electron microscopic view of basal body duplication shows two old bbs in cross-section and two new bbs (thick arrows), still in orthogonal position, anterior to the old bbs and aligned along the row. The A-P and R-L polarities of the rows are indicated: these polarities are indicated by the position and orientation of two basal body appendages, the ciliary rootlets (thin arrows)and a microtubule ribbon (arrowheads). Bar: 100 nm. Image: courtesy of N. Garreau de Loubresse.

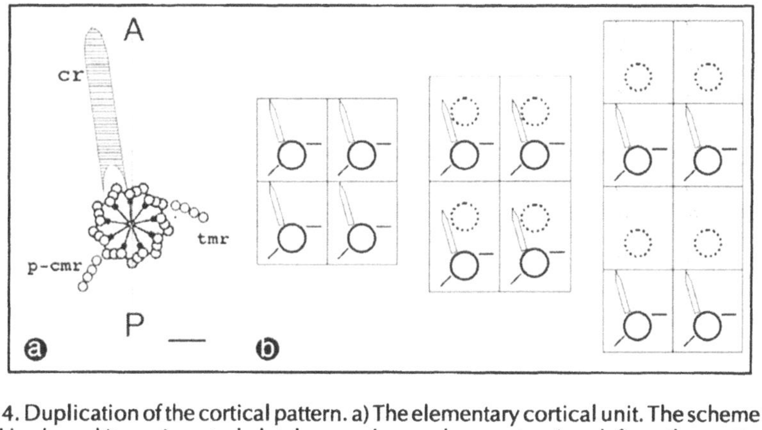

Figure 4. Duplication of the cortical pattern. a) The elementary cortical unit. The scheme shows a basal body and its main cytoskeletal appendages whose extension defines the cortical unit. The striated ciliary rootlet(cr) runs anterior and to the right of the basal body and of the cell; the transverse microtubule ribbon (tmr) and the post-ciliary microtubule ribbon (p-cmr) run to the left and the posterior sides of the basal body respectively. b) Duplication of the cortical units. Two adjacent units along two parallel rows are represented. Solid lines correspond to old bb and appendages, dotted lines to new organelles. Elongation of the longitudinal rows proceeds through elongation of the preexisting units and intercalation of new bbs.

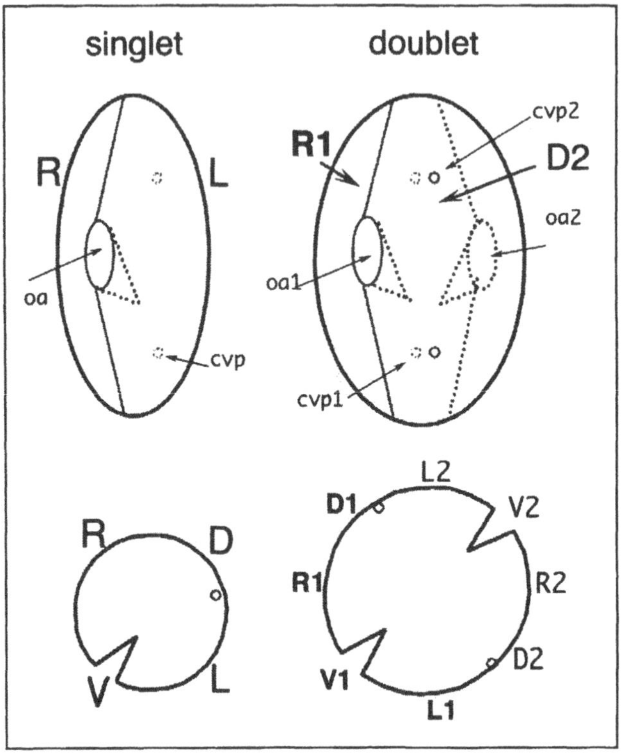

Figure 5. Organization of *Paramecium* doublet cells. The upper drawings outline a singlet and a homopolar doublet. In the singlet, the oral apparatus (oa) faces the observer and contractile vacuole pores (cvp) are on the hidden face. In the doublet, the second oral apparatus (oa2) is on the hidden side, while the corresponding contractile vacuole pores (cvp2) face the observer. The lower drawings represent the corresponding equatorial cross-sections, allowing a better visualization of the tandem organization of the doublet. R, L, V, D mark right, left, ventral, and dorsal sides of the singlet, respectively; R1/R2, V1/V2 etc. localize the different fields of the fused cells in the doublet.

limits of the cortical unit (Fig. 4). Although orchestrated at the whole cell level and triggered by the same mitotic signals, basal body duplication and the correlative duplication of cortical units are managed locally, at the cortical unit level (Fig. 3).

The pioneer experimental study started with doublet cells. "Doublets" had long been observed to appear occasionally in cultures of different ciliate species, and their stability was noted. *Paramecium* doublets comprise two complete sets of organelles and basal body fields, arranged in tandem (Fig. 5). This doublet organization is perpetuated through vegetative division and Sonneborn explored the genetic determation of this stability.

One advantage of *Paramecium* for genetic studies lies in the fact that each pair of conjugants simultaneously achieves the two reciprocal crosses, as conjugation involves only nuclear exchange: each conjugant retains its cellular integrity and a stationary, "female", gametic nucleus which fuses with a migratory, "male", gametic nucleus provided by the partner. Each pair of conjugants thus yields two F1 clones of identical heterozygous nuclear genotype but each retains its cytoplasmic characters. Occasional cytoplasmic bridges allow cytoplasmic exchanges, which can be monitored by cytoplasmically controlled characters such as the presence or absence of "kappa particles", a bacterial endosymbiont, or the alternative expression of mating types.[5] Doublets arise precisely when such bridges between two conjugants fail to break, leading to complete fusion of the two conjugants.

Doublets could be crossed with normal partners carrying genetic nuclear and cytoplasmic markers. The doublet phenotype was maternally inherited and maintained in F2 and subsequent sexual generations in the maternal line of descent, regardless of the segregation of the genetic markers, and even after intermixing of endoplasm and cytoplasmic organelles, thus ruling out both nuclear and cytoplasmic determinants. Still, it could be argued that the singlet/doublet alternative might reflect alternative differentiated states of the macronucleus (a process known to control expression of *Paramecium* surface antigens and mating type) or that doublets might have a double gene complement. Both counter-arguments could be ruled out by sophisticated manipulations amounting to reciprocal transplantations of nuclei.[1]

If the hereditary determinant for the doublet condition was neither in the nuclear genes nor in endoplasmic "information", its basis ought to reside in the cortex itself which would thus ensure its own perpetuation. This "cortical inheritance" suggested the more general concept of "cytotaxis" or "structural inheritance", namely the directive role of preexisting structures and organization on assembly and organization of new structures.[6]

Structural Inheritance in Ciliates: Theme and Variations

The study of more discrete cortical mutants provided some insight into the mechanisms underlying cortical inheritance. Variant cells were spotted by their abnormal "twisty" swimming, a clonally inherited property which turned out to reflect the presence of a complete ciliary row with 180° reversed polarity as judged by the orientation of the ciliary rootlets along this row (Fig. 6). In order to understand the origin and stability of this discrete phenotypic change, similar changes were experimentally generated by grafting a small piece of cortex in reverse polarity. Cytological analysis of a number of such "grafted cells" showed: (1) soon after surgery, the presence, as expected, of a few intercalated short segments of ciliary rows with reversed polarity; (2) the progressive elongation of these intercalated segments in the course of the next two-three divisions; (3) the establishment of complete intercalated ciliary rows of reverse polarity in sub-clones of the parent cell derived from the posterior product of the first division following the graft. All these lines manifested the "twisty" swimming behaviour, all the more pronounced as the number of inverted rows was greater.[7]

Cell lines with 2-3 to up to 12 inverted rows (among the ca. 70 normal longitudinal rows) were obtained. These cell lines retained their inverted rows for hundreds of cell divisions, with occasional loss of inverted rows. As in the case of doublets, genetic analysis ruled out any genetic change. The stable integration of the inverted rows in the global cortical pattern is all the more striking as it introduces a functional defect since, in the inverted rows, the cilia beat in the direction opposite to that of normally oriented cilia during both forward and backward swimming.[8] However, this defect, responsible for the twisty swimming behaviour, is not strongly counter-selected under laboratory conditions.

The simple geometry of this type of variation revealed the mechanism of its establishment and maintenance, which directly reflects the autonomous mode of duplication of basal bodies and cortical units. They carry and transmit information not only for assembly of organelles, but also for their polarity which thus appears not to be directly genetically determined. The morphogenetic autonomy of basal bodies and cortical units, the building blocks of the cortex, then also accounts for the cortical inheritance of the doublet condition.

An obvious question is: might these hereditary variations reflect some specificity of *Paramecium*, owing to its highly constrained mode of division, with strict conservation of all existing cortical structures and formation of new organelles, oral apparatus and contractile vacuole pores, close to preexisting cortical structures? Among the wide world of ciliates, species display different morphogenetic strategies and differ in regulative capacities. Thus "*Paramecium* is a ciliate counterpart to a mosaic embryo, whereas *Tetrahymena* is more like a regulative embryo".[9]

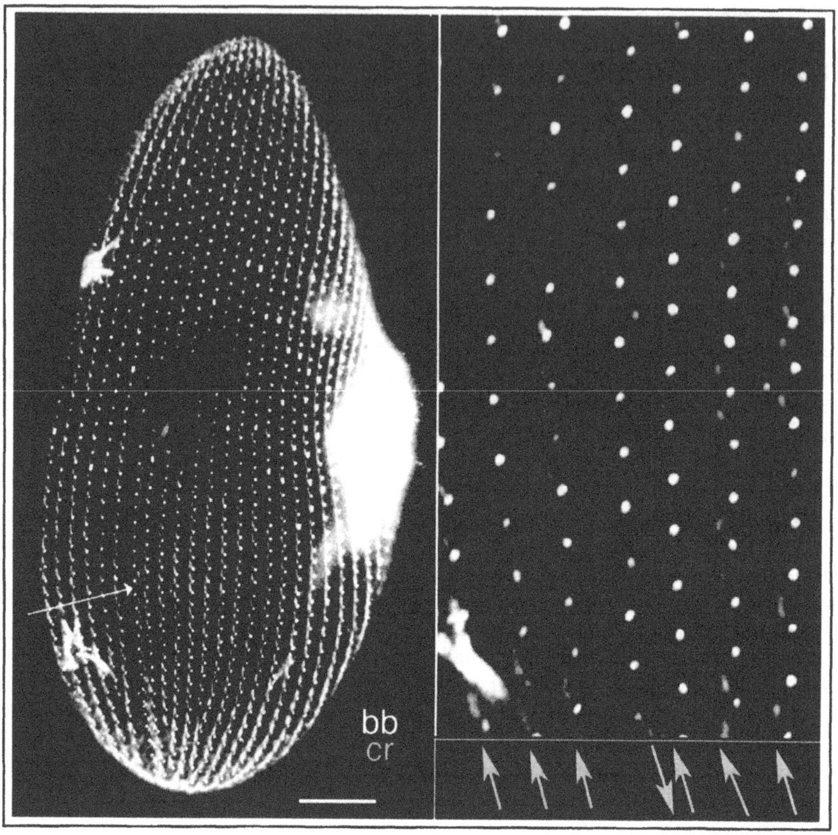

Figure 6. Visualization of an inverted ciliary row. As in Figure 1, double staining of basal bodies and ciliary rootlets indicates the polarity of basal bodies and cortical units. On the left panel, the arrow points to a discontinuity in the cortical pattern. On the right panel, the enlargement around this discontinuity shows the direction of the ciliary rootlets, which run downward and to the right of the observer in the case of the inverted row, while they run upward and to the left of the observer for normally oriented rows. Bar: 10 µm. Image: courtesy of F. Ruiz.

Nevertheless, 180° inverted rows were produced in *Tetrahymena* by the same type of micro-surgical method as in *Paramecium* and their hereditary maintenance demonstrated.[10] Both in *Tetrahymena* and in a related species, *Glaucoma*, further variant cortical configurations—mirror-image doublets and singlets displaying a reversal of cellular handedness relative to the antero-posterior cell axis—were also shown to be clonally inherited.[11,12]

Spirotrich ciliates like *Oxytricha* are still more highly regulative. They dedifferentiate and redifferentiate their ciliature at each division and can encyst, then loosing any ultrastructurally identifiable remnant of ciliature-associated stuctures.[13] However, upon excystment, they will redevelop their exact pre-cyst ciliature. So did doublets, whether homopolar or heteropolar, or cells possessing supernumerary dorsal structures: they all recapitulated their pre-cyst organization upon excystment.[14] Although in these case, genetic analysis was limited to homopolar doublets, the mode of origin and reversion of these variants indicated a non-genetic determinism.[14]

Beyond evidence for a structural memory what do these experiments demonstrate? Firstly, no one to one correspondence links genotype and cell architecture since the genotype of a cilate can stably accomodate a range of variations departing from the wild type. Secondly, and this was the more provocative aspect, the guiding role of the existing organization on the assembly of new structures implies transmission of a new type of information, non DNA-based. The term of "protein-based inheritance" was not then used, but it was pointed out that the presence of nucleic acids in basal bodies or any component of the cortex could not account for the observed phenomena, which do not bear on the properties of basal bodies or associated structures, but on variations in the spatial relationships among these elementary structures which themselves appear molecularly unaltered.[7]

Structural Inheritance and Basal-Body Biogenesis

At least part of the determinism of cortical inheritance depends on the mechanism of basal body duplication. It happens that the centriolar structure, namely the centriole or basal body, is the only cellular entity besides chromosomes which duplicates, and typically does so once per cell cycle. The underlying molecular mechanisms are not yet known, but clearly the process itself relies on structural inheritance, since in most cases, no such structure develops in the absence of a preexisting one. Only under particular biological conditions can a centriolar structure be formed de novo, for example in the development of the flagellated gametes of certain plants, in the flagellate form of the amoebo-flagellate *Naegleria* or during differentiation of ciliated epithelia in metazoa.[15-17] However none of these de novo formed basal bodies ever duplicate. Experimental suppression of existing basal bodies in *Chlamydomonas* or of centrioles in HeLa cells led to activation of the de novo pathway;[18,19] however it has not been established whether these de novo formed organelles were fully competent. The prevalent notion is that the two pathways have been conserved and coexist, but that, in presence of a preexisting structure, the templated pathway prevails as "kinetically dominant" and represses the alternative mode.

It is remarkable that the centriolar structure with its templated mode of duplication, which represents the paradigm of structural inheritance, appeared with the first eukaryote and has been totally conserved throughout evolution. In addition to its role as organizer of the cytoskeleton and cell architecture, there are several reasons for this conservation. First, the templated pathway provides a facilitated mode of assembly; secondly, it implies conservation of number, a crucial parameter in cellular division; last but not least, it ensures transmission of polarities through cell divisions, a function which is conspicuous in ciliates or flagellates and is likely to accompany duplication of centrioles (for a review see ref. 20). In fungi, all the functions of centrioles are fulfilled by a different organelle, the spindle pole body whose duplication, by direct budding at a preformed site of the mother organelle, also depends on structural inheritance and transmits polarities from mother cell to the bud.

Since an important function of centriolar structures is to nucleate a cilium or a flagellum, cilia and flagella might also be expected to transmit polarities since their circumferential polarities reproduce those of the basal body.[20] In a remarkable instance, described in trypanosome, the transmission of cell shape and polarity is precisely mediated by a direct connection between the developping new flagellum and the old one.[21,22] In metazoa, a role of primary cilia in the establishement of asymmetries and polarities during development is beginning to be recognized, as for determination of right-left asymmetry or for the functioning of the hedgehog signaling cascade regulating growth and patterning.[23-25]

Finally, there may be more to basal body duplication than the organelle itself and its appendages, and structural inheritance may turn out to tell a russian doll story. As noted above, the cysts of the spirotrich *Oxytricha* retain a blueprint of their cortical pattern. So does another spirotrich, *Paraurostyla*, and the cysts it forms during sexual processes, called zygocysts. An antibody raised against mammalian centrosomes was found to decorate, in the zygocyst, a transient system of tracks along which basal bodies redistribute upon excystment.[26] A recent study of *Chlamydomonas* reveals the existence of a filamentous centrin-containing scaffold that extends from the nucleus to the flagellar bases.[27] Such a centrin-based transcellular scaffold might contribute to embody the cellular geometry and might underly the well established role of these Ca^{2+}-binding proteins in duplication of centrioles, basal-bodies and spindle pole bodies.[28] The idea would also fit results obtained in *Paramecium*: functional analysis by RNAi showed that basal-body-associated centrins control the geometry of basal body duplication, i.e., the site of assembly of the daughter basal body.[29] Centrin (or a centrin-containing scaffold associated with the basal body) has recently been shown to be required for duplication and positioning of the Golgi stack in Trypanosome.[30] Such scaffolds subtending the cell surface might have a general role in transmission of polarities, not only in cells, but also through development. A striking example of predetermined polarity was demonstrated in the development of the quail oviduct, where assembly of ciliary basal bodies represent the last step of differentiation: in a 180° inverted segment of oviduct *primordium* replaced in situ, basal bodies differentiate and grow cilia that beat with reversed polarity.[31]

Inheritance of polarities—possibly mediated by a fibrillar scaffold—thus appears as a key parameter in cell memory with centriolar structures as major relays.

Structural Inheritance and the Limits of Direct Gene Control

The existence of structural inheritance challenges the idea that cellular organization is under direct gene control. This notion, which is a mere extension of the central dogma, implies that the properties of the proteins themselves are sufficient for their "self-assembly" into high order subcelullar structures, such as cytoskeletal organelles and networks. In fact, the concept of self-assembly, where self-assembled proteins reach a stable equilibrium, as in the case of a virus capsid assembly, cannot account for the dynamic properties of cytoskeletal structures, which never reach a stable equilibrium. To accommodate the properties of versatile cytoskeletal networks in vivo, the concept of "self-organization" describes a stable albeit dynamic structure.[32,33] However, if the specificities of the constitutive proteins and the dynamic properties of "self-organized" cytoskeletal polymers can generate both intrinsic polarity and pattern diversity, they still do not account for the 3D-deployment of the networks within the cellular space. Additional information intervenes: a seed or prepattern for nucleation and a (generally membranous) capture site or anchor which fulfills a "conformer" function and achieves a "directed assembly". In centriole/basal body duplication, a prepattern for the ninefold symmetry clearly preexists assembly of the microtubule cylinder (Fig. 3). Even the de novo pathway of assembly seems to involve, as a first step, assembly of a template resembling a cartwheel that provides the ninefold symmetry pattern.[17]

At the cellular level, the organization of the most general cytoskeletal constituent, microtubule networks, is determined by the existence of microtubule organizing centers (MTOCs) which provide focal sites of microtubule nucleation. The spatial deployment of the microtubules can be explained by the "dynamic instability" of these polymers: switches between phases of growth and shrinkage allow random patterns to become fixed by stabilisation at distal capture sites. This process of "dynamic instability" allows the network to display more than one organization but also an organization which is not directly defined by genetic information, but determined by the position of the MTOC and that of distal preexisting spatial cues, at kinetochores, membranes or cortex.[34] These degrees of freedom in pattern details mark the limits of gene control and underscore the determining role of preexisting spatial cues.

Even less dynamic, permanent, networks escape direct gene control of their pattern. An interesting example is again offered by *Paramecium*. The infraciliary lattice (ICL) is the innermost cytoskeletal network that subtends the cortex all over the cell: it forms a continuous net whose polygonal meshes, made of bundles of microfilaments, run around the bases of basal bodies (Fig. 7). The ICL is a contractile network, composed of numerous different centrins and of centrin-binding proteins.[35] Upon Ca^{2+} entry, this massive array of high-affinity Ca^{2+}-binding proteins causes cell contraction and correlative buffering of the deleterious action of the Ca^{2+} ions. Under laboratory conditions, the network is dispensable: inactivation of any of the centrin genes specific to the network leads to its disassembly, with no other effect on the cells than loss of contraction under a mild Ca^{2+}-stress.[36] In stably ICL-less cells, dots of centrin-containing material (which appear ultrastructurally amorphous in contrast to ICl meshes) are retained

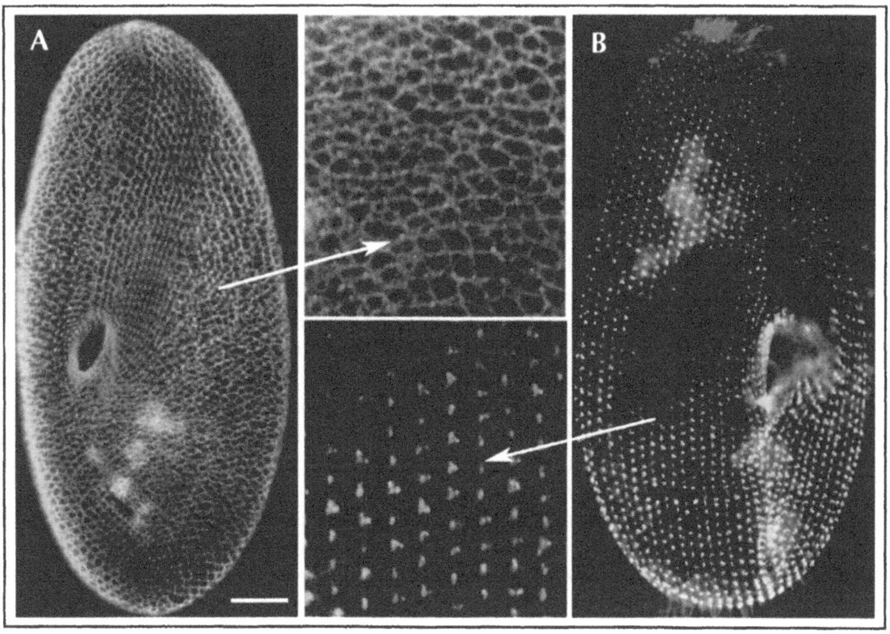

Figure 7. Organization of the infraciliary lattice. This cytoskeletal network is composed of centrins and centrin-binding proteins associated in bundles of microfilaments. The lattice is labeled by anticentrin antibodies (in green). A) In a wild type cell, the ICL network subtends the whole cell and each mesh encircles a basal body (labeled in red by an antitubulin antibody). B) In cells depleted of one of the constitutive centrins by RNAi treatment, the network is disassembled, but cortical units retain dots of anti-centrin reactive material flanking the basal bodies. Bar: 10 μm.

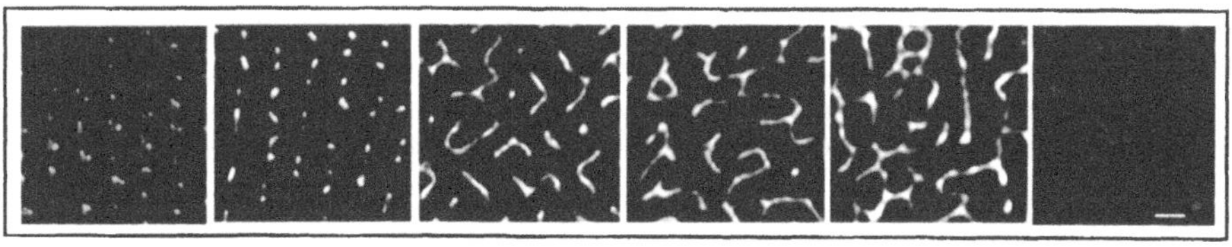

Figure 8. Mode of reassembly of the infraciliary lattice. Upon release of the silencing conditions, filaments arise from the residual dots flanking the basal body. The four images in black end white show, from left to right, the progress of reassembly over 1-3 divisions; and the network does not recover its normal pattern (rightmost image) until 10-15 divisions. Bar: 2 μm.

within each cortical unit (Fig. 7). These dots always flank the same side of the basal body, and are "reproduced" with basal bodies, since they are consistently present next to basal bodies in the course of cell divisions. Reactivation of the silenced gene(s) leads to ICL reassembly. The time course of the process (Fig. 8) shows that filament bundles are initiated only at the centrin-containing dots which we have called ICLOCs (for ICL-organizing centres), whose molecular components, besides centrins, are yet uncharacterized. The reassembly process presents two remarkable features: (1) at random places over the cell, short filaments emerge from an ICLOC; and (2) their elongation follows a random path until they merge into a continuum that eventually is stabilized into a pattern resembling the original one.[36] As for the organization of a microtubule array, the pattern of the ICL is not genetically programmed, but globally constrained by external preexisting spatial cues, and in particular by the position of the ICLOCs, which in turn reflects the basal body pattern... another Russian doll story.

The Tool Box of Structural Inheritance

Although we do not know how the ICL fibers are nucleated by the ICLOCs, the regeneration of the ICL in *Paramecium* reveals that self-sustained specialized organizing centers exist for cytoskeletal components other than the ubiquitous microtubules. But even to nucleate microtubules, cells have other resources than the trio of canonic MTOCs i.e., centrosome, spindle pole body and basal body. A simple device was first described in trypanosome where very short microtubules intercalate in the preexisting cortical array, as if nucleated by direct lateral interaction with existing adjacent microtubules, leading to a semi-conservative reproduction of the pattern.[37] A similar mechanism operates in plant cells where the extant cortical microtubules themselves bind γ-tubulin leading to a microtubule-dependent microtubule nucleation process which maintains the global organization and polarity of the cortical microtubule layer through longitudinal cell growth and division.[38] In *Paramecium*, devoid of centrosomes, aggregates of short microtubules and γ-tubulin complexes reside permanently within the germline micronuclear compartment, where they partition at division to organize the mitotic spindle.[39]

If centriolar structures and other self-reproducing nucleation centers represent the basic tools of structural inheritance, perpetuation of cell organization also resorts to an array of devices which do not involve self-reproducible organelles or molecular complexes but a diversity of more or less stable or transitory relays which nevertheless contribute to ordering of new by old cell structure/polarity across division. In *Drosophila* during ovogenesis, a special membranar structure, the fusome, present only over a small number of cell divisions, relays polarity from cystoblast to oocyte.[40] Similar situations can also be found in a "simple " unicell, *S. cerevisiae*. A well dissected polarity cue which marks the bud site and determines the axis of cell growth involves in particular Rax1p and Rax2p: these integral membrane proteins remain stably localized at the budding site through multiple cell cycles where they interact with the other proteins involved in the budding process.[41,42] Large immobile protein complexes, called eisosomes, have recently been described; they comprise both cytoplasmic and membrane proteins and their stability is thought to be important for maintaining organization of the plasma membrane.[43] More generally, during cell division and development, if molecular traffic is controlled upstream by gene encoded zipcodes in messenger RNAs or proteins, it is eventually dependent upon preexisting spatial cues and anchors. Finally, it may not be fortuitous that spatial cues are most often localized at the cell surface, that is at membranes which never form de novo.

Structural Inheritance as Protein Based-Inheritance

What is finally the fundamental meaning of the cortical inheritance demonstrated in ciliates and of the derived concept of structural inheritance? As first pointed out by Grimes and Aufderheide, it is basically of the same nature as prions which show that the same gene product can exist under two (or more) conformations whose accidental shift leads to new infectious and cytoplasmically transmissible phenotypes.[44] In structural inheritance, the observed variations do not affect protein conformation but rather the spatial arrangement of subcellular structures. However, in both types of phenomena, key steps involve nucleation processes and assembly of fibrous structures. At both molecular and subcellular levels, conformation is therefore not univocally determined by genetic information and the heritability of novel conformations of stochastic occurence "widens the phenotypic space encoded by a given genome... and can modify the adaptative landscape and influence evolution".[45] Considering that ciliary/flagellar basal bodies appeared with the first eukaryotes, it is hard to imagine that cortical variants with loss, addition or inversion of basal body territories may not have contributed at least to protist diversity.

From a historical point of view, it is interesting to note that when the initial work on *Paramecium* was published (1963-1965), the concept of structural inheritance had little impact and the cortical variants were often viewed as a zoological curiosity of limited significance. In fact, even though it was buttressed by a convincing genetic analysis, the demonstration that preformed cell structure played a key role in cell heredity encountered the same type of ideological barrier as the recognition of the "protein-only" status of prions. This experimental evidence for a protein-based cell memory was rarely integrated in biological research and this may in part explain why so few recognized cases of structural inheritance have been reported or explored: molecular or structural landmarks which provide evidence of even a short-term cell memory in the course of differentiation and development are not perceived as such but as a particular step along a linear sequence of reactions starting from the genes.

With the renewal of interest for epigenetic phenomena, and the unlimited possibilities offered by all the new "omics" and tools for functional analysis, many more landmarks, anchors, and structural relays involved in division and development will be characterized, and reveal the molecular circuitry underlying cell memory.

Acknowledgements

I would like to thank all the Gif Paramecium team, and in particular Jean Cohen, Linda Sperling and Françoise Ruiz for our daily stimulating discussions. I gratefully acknowledge the continuous support of Paramecium research by the Centre National de la Recherche Scientifique.

References

1. Sonneborn TM. Does preformed cell structure play an essential role in cell heredity? In: Allen JM, ed. The Nature of Biological Diversity. New York: McGraw-Hill, 1963:165-222.
2. Griffith JS. Self-replication and scrapie. Nature 1967; 215:1043-1044.
3. Iftode F, Fleury-Aubusson A. Structural inheritance in Paramecium: Ultrastructural evidence for basal body and associated rootlets polarity transmission through binary fission. Biol Cell 2003; 95:39-51.
4. Dippell RV. The development of basal bodies in Paramecium. Proc Natl Acad Sci USA 1968; 61:461-468.
5. Sonneborn TM. Partners of the genes. Scient Amer 1950; 30-39.
6. Sonneborn TM. The Differentiation of Cells. Proc Natl Acad Sci US 1964; 51:915-929.
7. Beisson J, Sonneborn TM. Cytoplasmic Inheritance of the Organization of the Cell Cortex in Paramecium aurelia. Proc Natl Acad Sci US 1965; 53:275-282.
8. Tamm SL, Sonneborn TM, Dippell RV. The role of cortical orientation in the control of the direction of ciliary beat in Paramecium. J Cell Biol 1975; 64:98-112.
9. Frankel J. Structural inheritance. Pattern Formation: Ciliate studies and models. Oxford: Oxford University Press, 1989: 69-93.
10. Ng SF, Frankel J. 180° rotation of ciliary rows and its morphogenetic implications in Tetrahymena pyriformis. Proc Natl Acad Sci USA 1977; 74:1115-1119.
11. Nelsen EM, Frankel J, Jenkins LM. Nongenic inheritance of cellular handedness. Development 1989; 105:447-456.
12. Suhama M. Reproducing singlets with an inverted oral apparatus in Glaucoma scintillans (Ciliophora, Hymenostomatida). J Protozool 1985; 32:454-459.
13. Grimes GW. Morphological discontinuity of kinetosomes during the life cycle of Oxytricha fallax. J Cell Biol 1973; 57:229-232.
14. Hammersmith RL, Grimes GW. Effects of cystment on cells of Oxytricha fallax possessing supernumerary dorsal bristle rows. J Embryol Exp Morphol 1981; 63:17-27.
15. Wolniak SM, Klink VP, Hart PE et al. Control of development and motility in the spermatozoids of lower plants. Gravit Space Biol Bull 2000; 13:85-93.
16. Fulton C, Dingle AD. Basal bodies, but not centrioles, in Naegleria. J Cell Biol 1971; 51:826-836.
17. Anderson RG, Brenner RM. The formation of basal bodies (centrioles) in the Rhesus monkey oviduct. J Cell Biol 1971; 50:10-34.
18. Marshall WF, Vucica Y, Rosenbaum JL. Kinetics and regulation of de novo centriole assembly: Implications for the mechanism of centriole duplication. Curr Biol 2001; 11:308-317.
19. La Terra S, English CN, Hergert P et al. The de novo centriole assembly pathway in HeLa cells: Cell cycle progression and centriole assembly/maturation. J Cell Biol 2005; 168:713-722.

20. Beisson J, Jerka-Dziadosz M. Polarities of the centriolar structure: Morphogenetic consequences. Biol Cell 1999; 91:367-378.
21. Moreira-Leite FF, Sherwin T, Kohl L et al. A trypanosome structure involved in transmitting cytoplasmic information during cell division. Science 2001; 294:610-612.
22. Briggs LJ, McKean PG, Baines A et al. The flagella connector of Trypanosoma brucei: An unusual mobile transmembrane junction. J Cell Sci 2004; 117:1641-1651.
23. Nonaka S, Tanaka Y, Okada Y et al. Randomization of left-right asymmetry due to loss of nodal cilia generating leftward flow of extraembryonic fluid in mice lacking KIF3B motor protein. Cell 1998; 95:829-837.
24. Huangfu D, Anderson KV. Cilia and Hedgehog responsiveness in the mouse. Proc Natl Acad Sci USA 2005; 102:11325-11330.
25. Beales PL. Hedgehogs on the road to polarity. Nat Genet 2006; 38:277-279.
26. Fleury A, Le Guyader H, Iftode F et al. A scaffold for basal body patterning revealed by a monoclonal antibody in the hypotrich ciliate Paraurostyla weissei. Dev Biol 1993; 157:285-302.
27. Geimer S, Melkonian M. Centrin scaffold in Chlamydomonas reinhardtii revealed by immunoelectron microscopy. Eukaryot Cell 2005; 4:1253-1263.
28. Rice LM, Agard DA. Centriole duplication: Centrin in on answers? Curr Biol 2002; 12:R618-619.
29. Ruiz F, Garreau de Loubresse N, Klotz C et al. Centrin deficiency in Paramecium affects the geometry of basal-body duplication. Curr Biol 2005; 15:2097-2106.
30. He CY, Pypaert M, Warren G. Golgi duplication in Trypanosoma brucei requires Centrin2. Science 2005; 310:1196-1198.
31. Boisvieux-Ulrich E, Sandoz D. Determination of ciliary polarity precedes differentiation in the epithelial cells of quail oviduct. Biol Cell 1991; 72:3-14.
32. Kirschner M, Gerhart J, Mitchison T. Molecular "vitalism". Cell 2000; 100:79-88.
33. Kushner DJ. Self-assembly of biological structures. Bacteriol Rev 1969; 33:302-345.
34. Kirschner M, Mitchison T. Beyond self-assembly: From microtubules to morphogenesis. Cell 1986; 45:329-342.
35. Klotz C, Garreau de Loubresse N, Ruiz F et al. Genetic evidence for a role of centrin-associated proteins in the organization and dynamics of the infraciliary lattice in Paramecium. Cell Motil Cytoskeleton 1997; 38:172-186.
36. Beisson J, Clerot JC, Fleury-Aubusson A et al. Basal body-associated nucleation center for the centrin-based cortical cytoskeletal network in Paramecium. Protist 2001; 152:339-354.
37. Sherwin T, Gull K. Visualization of detyrosination along single microtubules reveals novel mechanisms of assembly during cytoskeletal duplication in trypanosomes. Cell 1989; 57:211-221.
38. Murata T, Sonobe S, Baskin TI et al. Microtubule-dependent microtubule nucleation based on recruitment of gamma-tubulin in higher plants. Nat Cell Biol 2005; 7:961-968.
39. Klotz C, Ruiz F, Garreau de Loubresse N et al. Gamma-tubulin and MTOCs in Paramecium. Protist 2003; 154:193-209.
40. Huynh JR, St Johnston D. The origin of asymmetry: Early polarisation of the Drosophila germline cyst and oocyte. Curr Biol 2004; 14:R438-449.
41. Kang PJ, Angerman E, Nakashima K et al. Interactions among Rax1p, Rax2p, Bud8p, and Bud9p in marking cortical sites for bipolar bud-site selection in yeast. Mol Biol Cell 2004; 15:5145-5157.
42. Chen T, Hiroko T, Chaudhuri A et al. Multigenerational cortical inheritance of the Rax2 protein in orienting polarity and division in yeast. Science 2000; 290:1975-1978.
43. Walther TC, Brickner JH, Aguilar PS et al. Eisosomes mark static sites of endocytosis. Nature 2006; 439:998-1003.
44. Grimes G, Aufderheide KJ. Cllular aspects of pattern formation: The problem of assembly. Karger (Basel), 1991.
45. Shorter J, Lindquist S. Prions as adaptive conduits of memory and inheritance. Nat Rev Genet 2005; 6:435-450.

Chapter 10

Centriole Inheritance

Patricia G. Wilson*

Abstract

Early cell biologists perceived centrosomes to be permanent cellular structures. Centrosomes were observed to reproduce once each cycle and to orchestrate assembly of a transient mitotic apparatus that segregated chromosomes and a centrosome to each daughter at the completion of cell division. The molecular picture of centrosomes is now more complete. Centrosomes are composed of a pair of centrioles buried in a complex pericentriolar matrix. The bulk of microtubules in cells lie with one end buried in the pericentriolar matrix and the other extending outward into the cytoplasm. Centrioles recruit and organize pericentriolar material. As a result, centrioles dominate microtubule organization and spindle assembly in cells born with centrosomes. Centriole inheritance is ensured by cell cycle checkpoints that monitor the presence of centrosomes in G1 and spindle bipolarity in mitosis. In spite of these cell cycle controls, assembly of bipolar spindles in acentriolar cell extracts and cell proliferation in acentriolar mutants argue that neither centrioles nor centrosomes are essential structures. This review examines the recent advances in understanding centriole inheritance.

Introduction

Theodor Boveri (1862-1915) was among the first biologists to recognize the importance of centrosomes to inheritance of genetic information.[1,2] Boveri and his contemporaries accepted that cells divide by binary fission and that chromosomes provide the genetic basis of inheritance. How the thread like chromosomes in the nucleus of the parental cell are transmitted to the two daughter cells was a mystery. Through careful observations with the crudest of microscopes by today's standards, but the keenest of insights, Boveri defined the centrosome (Fig. 1) as a pair of centrioles surrounded by a special material (pericentriolar matrix) that organized the surrounding cytoplasm and formed a "division center" (microtubule aster). Boveri argued that the centrosome was the "reproductive organ" of the cell, orchestrating assembly of a transient mitotic apparatus (spindle) that segregated bipartite chromosomes (sister chromatids) to each daughter cell. Central to his argument, centrosomes are permanent cellular structures that have the capacity to reproduce and divide once per cell cycle. He used simple tricks to generate cells with one or more centrosomes, finding a monopolar half-spindle organized by a single centrosome and multipolar spindles organized by a centrosome at each pole. He argued that the number of centrosomes determined the number of spindle poles. The Centrosome Theory formulated by Boveri met resistance; not all cells have centrosomes, but assemble spindles nonetheless.

A landmark study provided proof of principle that centrosomes are not essential for spindle assembly.[3] Xenopus eggs lack centrioles, but contain all of the materials needed for many rounds of centrosome assembly. Chromatin coated beads in this system are substituted for chromosomes.

*Corresponding Author: Patricia G. Wilson—Regenerative Bioscience Center, Department of Animal and Dairy Science, University of Georgia, Athens, Georgia 30602 USA. Email: pgwilson2@earthlink.net

Protein-Based Inheritance, edited by Yury O. Chernoff. ©2007 Landes Bioscience.

Microtubules emerge from the chromatin coated beads, nucleated by members of a family of plus-end binding proteins. These proteins mediate tubulin addition at microtubule plus-ends, effectively pushing microtubule minus-ends outward. Microtubules are eventually bundled into spindle poles by microtubule dependent motor proteins. Spindle assembly in acentriolar Xenopus egg extracts suggests that cells have a default pathway to spindle assembly. The significance of these results are supported by studies showing that the components and molecular mechanisms that promote self-assembly of spindles in Xenopus extracts operate in vivo as well, nucleating microtubules that extend from the kinetochores of chromosomes and incorporate into spindle poles organized by centrosomes.[4,5] However, the central prediction of a default pathway is that monopolar spindles should be transformed into monastral bipolar structures in which only the astral pole is organized by centrosomes. Monastral bipolar spindles have been reported, but they result from direct[6-10] or indirect[11,12] effects on centriole assembly and/or centrosome organization that occurs subsequent to centrosome separation in prophase of mitosis and/or assembly of a bipolar spindle.[11,12]

Centrioles dominate microtubule organization and thus chromosome inheritance. Centrioles are the core of centrosome function and organization. Centrioles have a highly conserved 9 fold axis of symmetry (Fig. 1), most often formed by triplets of microtubules in a pinwheel. Centrioles accumulate pericentriolar material and as a result, centrioles effectively organize microtubules in a radial array. Centrioles are virtually identical to the core of basal bodies that organize the axonemes of cilia and flagella (Fig. 3). Defects in basal body function have profound effects on human health,[13] but space limitations preclude further discussion here. This review will examine the recent literature on the function, assembly, and inheritance of centrioles by mitosis in light of recent debate on the need, or lack thereof, for centrioles in mitosis.

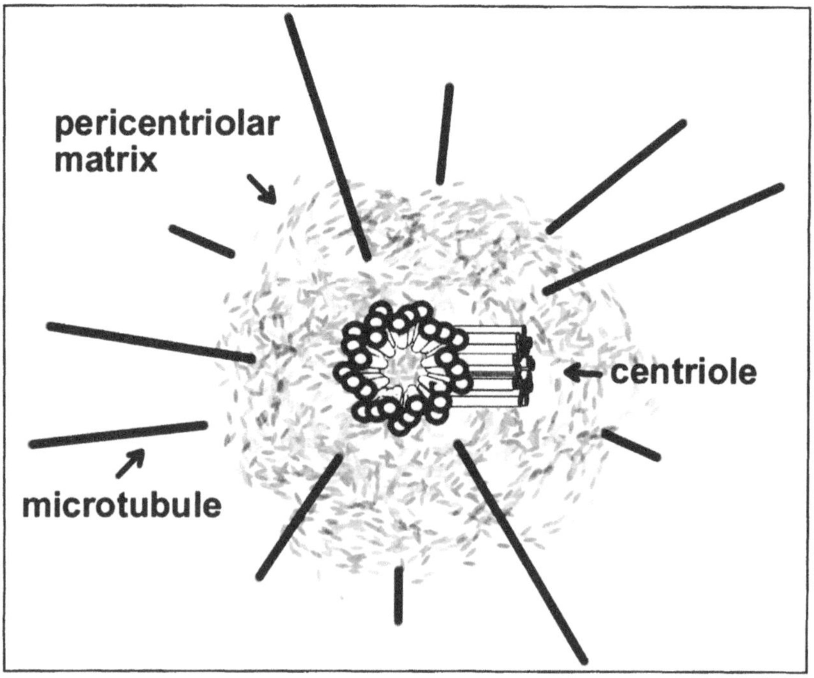

Figure 1. Centrosome structure. Pericentriolar material surrounds a pair of barrels-shaped centrioles. Minus ends of astral microtubules lie in pericentriolar matrix and plus ends extend outward.

Centriole Function in Mitosis

Microtubule Organizing Centers

Expectations of centrosome activity and the corresponding role ascribed to centrioles are rooted in microtubule dynamics. Microtubules are closed tubes of ~13 protofilaments of tubulin monomers, αβ-tubulin heterodimers, in head-to-tail polymers.[14] Tubulin monomers are held together by noncovalent interactions and as a result, microtubules are intrinsically dynamic and constantly turnover in cells, changing both in length and composition by addition and loss of tubulin at microtubule ends. The designated plus-end of microtubules is more dynamic, frequently growing or shrinking. The opposing minus end is also dynamic although tubulin is lost, but not added in vivo. Tubulins are GTP binding proteins and polymerization into growing microtubules requires that both α- and β-tubulins are bound to GTP. Crystallography of α-/β-tubulin bound to GTP shows a straight conformation that is thought to promote the longitudinal interactions necessary for polymerization.[15] GTP bound to β-tubulin subsequently undergoes hydrolysis, resulting in a more unstable lattice.[16] Tubulin self-assembles into microtubules in vitro; the rate at which new microtubules form, designated as nucleation, depends heavily on the concentration of tubulin, GTP and microtubule associated proteins (MAPs) that lower the concentration of tubulin needed for self-assembly of comparable amounts of polymer. Until recently the concentration of tubulin in cells was considered too low to support the level of tubulin polymer found in cells by self-assembly, requiring a mechanism to renew the microtubules lost to depolymerization. Because the bulk of microtubules is organized by centrosomes, microtubule nucleation was inferred to be the essential activity of centrosomes and its pericentriolar components.

The Pericentriolar Matrix

Among pericentriolar components, γ-tubulin has been the most extensively studied and is widely used as a direct or indirect metric of centrosome assembly and activity. Members of this family show more structural diversity than the highly conserved α- and β-tubulins, but γ-tubulin is also ubiquitously expressed among eukaryotes and essential for viability. Early studies raised the question of whether γ-tubulin interacts with α/β-tubulin in the same way that monomers of α/β-tubulin interact with each other during polymerization into protofilaments.[17] A portion of γ-tubulin in extracts of flies and vertebrates sediments in high molecular weight ring-like complexes, designated γ-TuRCs, with 5-6 other proteins (Table 1). The shape of γ-TuRCs is compelling, but stable small complexes (γ-TuSCs) with only 2 additional proteins are also found in Drosophila and these do not have a ring-like structure. Further, the components of γ-TuRCs are not expressed in all species (Table 1) and the results of recent crystallography studies are not easily reconciled with template activity. Crystals of γ-tubulin bound to a nonhydrolysable GTP analog showed a curved conformation that is most similar to α-/β-tubulins incorporated into the walls of microtubules and in soluble polymerization-incompetent monomers.[15] Localization to microtubule minus ends has received the most attention,[18] but γ-tubulin is periodically positioned along the length of microtubules in fission yeast[19] and in land plants where it is needed for assembly of branched cortical microtubule arrays.[20,21] While γ-tubulin containing complexes in the pericentriolar matrix are credited with nucleating microtubules at centrosomes, the contributions of these complexes are not yet clear.

Although initial studies suggested that γ-tubulin might be universally essential for microtubule nucleation, it is now widely acknowledged that alternative pathways exist, in large part because abundant microtubules persist in γ-tubulin deficient mutants.[10,23-26] Loss of γ-tubulin function disrupts centrosome organization and generates malformed mitotic spindles. Severe γ-tubulin mutants generate monopolar-like structures; during prophase of the first mitosis in γ-tubulin deficient nematodes, the two centrosomal asters "fly in" toward each other and the terminal phenotype is a monopolar structure.[25,26] Evidence of spindle fusion was detected in γ-tubulin deficient Drosophila embryos (Fig. 2).[24,27] It is not known why loss of γ-tubulin

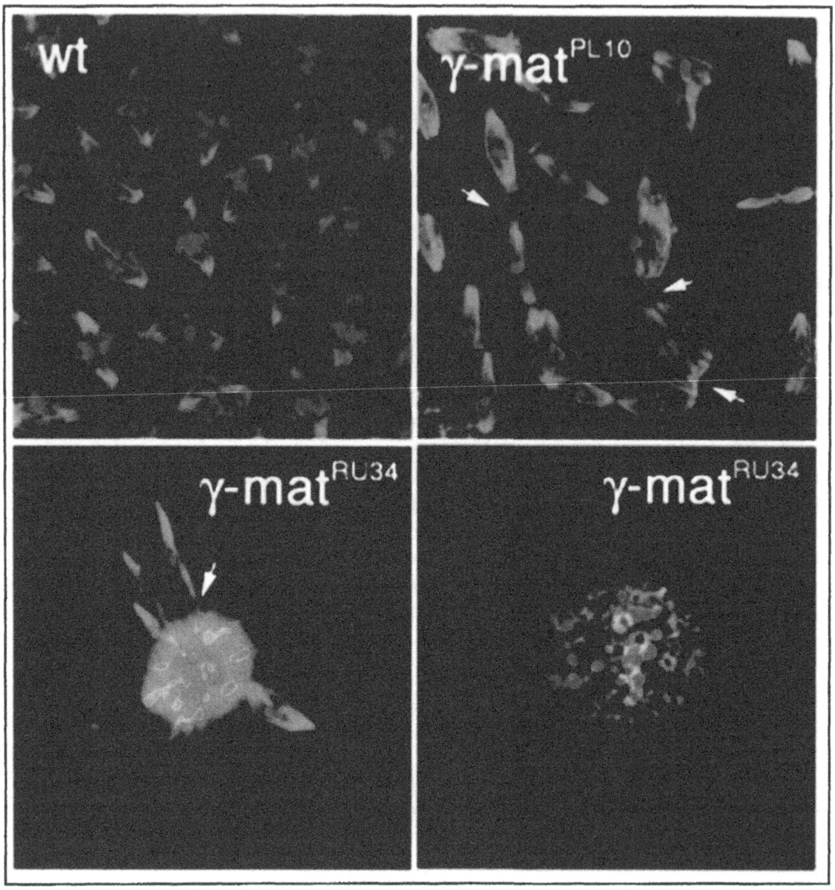

Figure 2. Spindle fusions in γ-tubulin mutants. Spindles in embryos from wild-type (wt) and γ-Tub37C (γ-mat) mutants as indicated. Embryos were stained with antibodies against α-tubulin to visualize microtubules and a fluorescent chromatin dye to visualize chromosomes. Arrows indicate some sites of apparent spindle fusion. Magnification identical in all panels.

function results in such fusion events. Inactivation of other γTuRC components has variable effects. The glp-1 protein is the only component of the γTuRC found in nematodes and the impact of its loss is similar to that of γ-tubulin. Inactivation of the γTuRC components Dgp75 and Dgp128 in Drosophila indicates these genes are not essential for viability. These mutants are viable and show only mild mitotic defects in somatic cells,[9] although more severe defects are generated in gametogenesis.[28,29] Inactivation of γ-tubulin, Dgp84 and Dgp71 by RNAi disrupt the spindle assembly checkpoint in cultured Drosophila cells and alter spindle pole organization.[10] On the basis of immunofluorescence, γ-tubulin is weakly associated with spindle microtubules in flies and vertebrates,[17] but it is not clear whether spindle localization is related to function in the spindle assembly checkpoint. The complex picture generated by genetic studies suggests that γ-tubulin and/or its cohorts have multiple functions, some of which are independent of centrosomes and microtubule organization.

Table 1. **γ*-Tubulin containing complexes*[a]**

	Humans	Frogs	Flies	Budding Yeast	Fission Yeast	Nematodes
γTuSC[b]	TUBG1	Xgam	γTub23C[c] γTub37C	Tub4	Gtb1	tgb-1
GCP2	Xgrip110	Dgp84	Spc97	Alp4		
GCP3	Xgrip109	Dgp91	Spc98	Alp6	glp-1	
γTuRC[d]	GCP4	Xgrip76	Dgp75	Gfh1		
GCP5	Xgrip133	Dgp128				
GCP6	Xgrip210	Dgp163	Alp16			
peripheral[e]	GCPWD	XNedd-1	Dgp71WD			

a. Components of γ-tubulin complexes have been reviewed recently.[17] b. Components corresponding to those in γTuSC are included in γTuRCs in vertebrates and flies.[17] c. Drosophila encodes two differentially expressed isoforms.[22] d. γTuRCs sediment from extracts of vertebrates. Both γTuRC and γTuSC sediment from extracts of fly embryos. e. Nonconserved peripheral component of γTuRCs in flies and vertebrates is required for localization of γ-tubulin to spindle microtubules.

Microtubule Asters

The number of microtubules organized by the pericentriolar matrix is commonly used as a measure of microtubule nucleation, but likely represents a composite of activities. If γ-tubulin is used as the metric, the mass of γ-tubulin at the onset of mitosis is ~ 3X that in interphase.[30] Accumulation of γ-tubulin is independent of microtubules and genetic studies indicate that long coiled coil proteins in Drosophila,[31] Caenorhabditis,[32] and vertebrates[33] are involved. Microtubule density in Caenorhabditis embryos was reduced in γ-tubulin depleted embryos to approximately 40% of wild-type levels by tubulin immunofluorescence,[26] suggesting that γ-tubulin acts in a kinetically dominant pathway to microtubule nucleation.[26] However, the kinetics of microtubule nucleation were not measured, but inferred. This is an important distinction because the pericentriolar matrix is complex and loss of one protein can cause loss of others.

RNAi inactivation of γ-tubulin in Caenorhabditis also reduces centrosome localization of ZYG-9,[26] an ortholog of XMAP215 in Xenopus. Loss of ZYG-9 reduced the number of centrosomal microtubules to a level comparable to loss of α-tubulin and more severely than loss of γ-tubulin.[34] Likewise, reconstitution of microtubule asters with salt stripped centrosomes requires XMAP215 in vertebrates.[35] Orthologs of XMAP215 are enriched at centrosomes, but

***Table 2. The extended tubulin superfamily*[a]**

	γ-tubulin	δ-tubulin	ε-tubulin	η-tubulin	ζ-tubulin
humans	+	+	+		
frogs	+	+	+		
flies	+				
nematodes	+				
ciliates	+	+	+	+	
trypanosomes	+	+	+		+

a. Additional tubulins identified by homology[43] are not listed here.

also bind along the length of microtubules and promote dramatic growth at microtubule plus-ends.[36-38] Remarkably, recombinant XMAP215 bound to beads nucleated microtubule asters in dilute solution of pure tubulin as well as in cell extracts.[35] In addition to γ-tubulin, glp-1 and ZYG-9, targeted RNAi screens in Caenorhabditis for genes involved in microtubule nucleation and growth implicate proteins known to interact with ZYG-9, proteins previously identified in other pathways and novel proteins that have not yet been fully analyzed.[34] Together with results from flies and vertebrates, these findings indicate that the density of microtubules at centrosomes likely reflects a composite of activities.

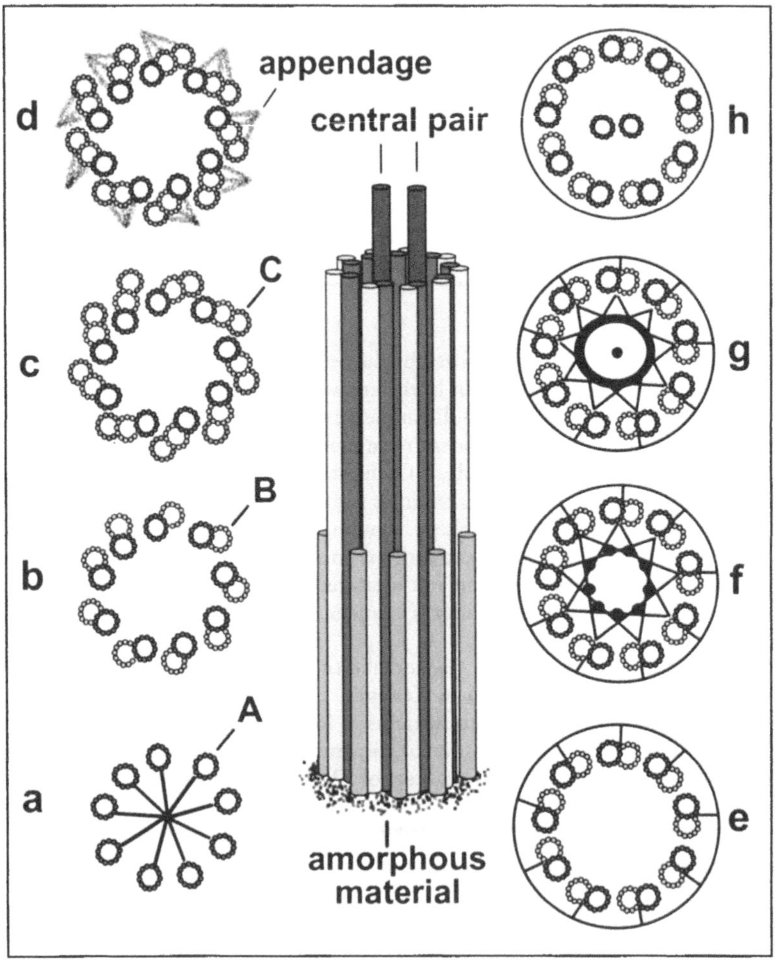

Figure 3. Centriole and basal body structure. Amorphous material is the first indication of a forming centriole or basal body. a-c. The A-, B- and C-tubules appear sequentially. d. Appendages form at the distal end of the parental centriole. e-g. The transition zone marks the end of C-tubules in basal bodies. B- and C -tubules extend into axonemes. h. The central pair of microtubules emerges in the transition zone, but they are not present in cilia. e-h. The plasma membrane surrounding basal body/axonemes is depicted as a circle. Modified from tomography of centriole/basal body structure in Chlamydomonas.[44]

Centriole Assembly and Acquisition of Pericentriolar Material

Centriole Function and Assembly

Centriole dominance in inheritance reflects the capacity of centrioles to organize pericentriolar components. The mechanism remains a mystery, but the effect is evident in the mouse blastocyst where centrioles do not appear until the second or third mitosis. The pericentriolar protein pericentrin[33] forms a broad band at acentriolar spindle poles, but the characteristic dense spherical focus was detected only after centrioles appear.[39] Likewise, γ-tubulin in acentriolar Xenopus egg extracts is distributed throughout acentriolar spindles rather than localized in a discrete focus at spindle poles.[40] An early study localized γ-tubulin to the pericentriolar material of centrosomes by ultrastructural analysis,[41] but another placed γ-tubulin in the centriole barrel.[42] Subsequent genetic studies revealed an essential role for γ-tubulin in centriole assembly as well as assembly of the central pair of microtubules that emerge from the transition zone in basal bodies (Fig. 3).[43] The precise role of γ-tubulin in centriole assembly is not known, however defects in centriole assembly likely contribute to formation of malformed spindles in γ-tubulin mutants. This view is consistent with differential effects on mitosis and female meiosis in γ-tubulin mutants in Drosophila; in contrast to mitotic spindles that are malformed, acentriolar meiotic spindles are bipolar in eggs lacking γ-tubulin or expressing an altered form.[24]

Almost a decade following discovery of γ-tubulin, the δ-,η-,ε-and ζ-tubulins were identified by genetic and/or genomic approaches.[43] In contrast to γ-tubulin, the δ-,η-,ε-and ζ-tubulins have limited distribution among species (Table 1). Although little is known about η-and ζ-tubulin function, basal bodies in Chlamydomonas and Paramecium mutants deficient in δ- and ε-tubulins are shortened and missing the B- and/or C-tubules and ε-tubulin has been implicated in centriole assembly.[43] The absence of δ and ε-tubulins in Caenorhabditis and Drosophila prompted models in which these tubulins are needed for assembly of the B- and C-tubules in centrioles.[45] This was founded on the presence of only the A-tubule in Caenorhabditis centrioles and the A- and B-tubules in centrioles of Drosophila embryos. Given that centrioles in the Drosophila germ lines have triplet microtubules, an alternative explanation is that δ- and ε-tubulins promote stability of centrioles/basal bodies. Stabilizing function is consistent with restored triplet microtubules in δ-tubulin deficient Chlamydomonas mutants that also express an altered form of α-tubulin. This mutant form of α-tubulin stabilizes microtubules and δ-tubulin may stabilize triplet microtubules in the same way.[43] While centrioles in whole Drosophila embryos contain doublet microtubules, centrioles in isolated centrosomes show singlet microtubules.[43] Both singlet and doublet microtubules could reflect the very rapid cell cycles in early Drosophila embryos of 8 to 20 minutes that may curtail centriole assembly without C-tubules, rendering the remaining B-tubule less stable in isolated centrosomes.

The Centriole Cycle and the Spindle Assembly Checkpoint

Centriole assembly is tightly controlled by the cell cycle. Each cell inherits a centriole pair at the completion of mitosis and cytokinesis of its parent (Fig. 4). Centrioles initially show an orthogonal orientation that is lost before assembly of a new centriole is initiated. A nascent centriole or procentriole appears during S-phase near the proximal end of the preexisting or parental centriole.[46] Initiation of centriole assembly in flies and vertebrates requires Plk-4 (polo kinase 4), a member of a conserved family of serine/threonine kinases. Plk-4 is centrosome associated throughout the cell cycle, but is activated near the G1/S phase transition by Cdk2-CyclinA/E phosphorylation.[47] Overexpression of Plk-4 in human HeLa cells generates several procentriole-like structures that encircle the proximal ends of the parental centriole.[7] These results suggest that multiple procentrioles can form in association with a preexisting centriole and that Plk-4 or its substrates may be rate limiting. However, the mechanism that limits assembly to a single centriole in normal cells is still a mystery. Procentrioles continue to elongate during G2, but typically do not acquire a significant mass of pericentriolar material or associated microtubules until the next cell cycle. Centrioles remain tethered in most cells until

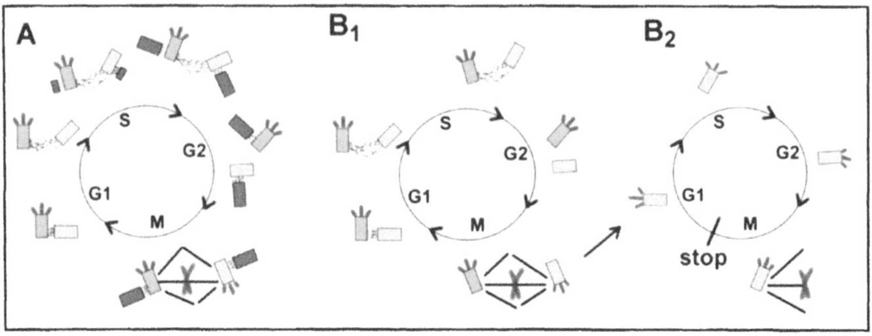

Figure 4. The centriole cycle and the spindle assembly checkpoint. A. Normal centriole cycle. B_1. Failure in centriole duplication, but centrosomes with a single centriole assemble a bipolar spindle. B_2. A single centrosome is partitioned to daughter cells that generates a monopolar spindle. Centriole pairs are connected by fibrillar material, designated by black dots. Parental centriole shows distal appendages (blue lines) that may be vestiges of centriole assembly since they are not essential for mitosis or viability.[48] Pericentriolar matrix and microtubule asters are not depicted for simplicity.

the G2/M transition into mitosis when the connection is severed, allowing centrosomes to separate and form a bipolar spindle.[46] If centriole duplication fails, the barren centrioles will separate and organize the poles of a bipolar spindle, but single centrioles will be partitioned by cytokinesis and these will organize a monopolar spindle in the following cell cycle (Fig. 4). In the absence of bipolar attachment of kinetochores, the spindle assembly checkpoint will arrest the cell cycle. In effect, the spindle assembly checkpoint serves as one backup mechanism to block proliferation of cells with defects in centriole duplication.

Interdependent Relationship Between Centrioles and the Pericentriolar Matrix

Localization of the pericentriolar component γ-tubulin to centriole barrels was one of the first molecular indications that centrioles and pericentriolar matrix are interdependent rather than functionally and structurally discrete. The relationship between centriole assembly and the pericentriolar matrix was brought into focus by genetic studies in Caenorhabditis. RNAi screens, chemical mutagenesis, and genomic approaches identified the novel zygotic defective ZYG-1 kinase, the spindle assembly proteins SAS-4, SAS-5, SAS-6 and the spindle defective protein SPD-2.[49] Results obtained with GFP-tagged proteins and mutant analysis revealed a dependent pathway in which localization of SPD-2 is the essential first step. The signal that elicits SPD-2 localization is not known, but SPD-2 appears before ZYG-1.[50] Caenorhabditis does not encode a Plk-4 homolog and it is not clear whether ZYG-1 provides the same or a different function. However, ZYG-1 activity is necessary for subsequent recruitment of SAS-5 and SAS-6, proteins that are mutually dependent for localization to centrioles and necessary for subsequent recruitment of SAS-4. Ultrastructural and immunolocalization results indicate that SAS-4, SAS-5, and SAS-6 are incorporated into the centriole barrel. The underlying mechanism is not understood, but SAS-5 continually cycles between centrioles and the cytoplasm while SAS-6 remains centriolar. Potential homologs in higher eukaryotes include SAS-6 and SAS-4 in humans and in flies as well as SPD-2 in humans. Functional analysis of human SPD-2 is not available and SAS-5 has not been found in other species, but the available data support a role for the human and fly homologs of SAS-4 and SAS-6 in centrosome assembly. A caveat to these results is that this set of SAS and SPD proteins contain coiled coil domains as do many centrosomal proteins. Coiled coil structure may mediate centrosome localization, but these proteins may not function the same way. That said, the highly conserved 9-fold symmetry of centrioles suggests assembly by a conserved core of proteins and activities that may include SAS and SPD counterparts.

Analysis of centrosome organization suggests that centriole assembly and accumulation of pericentriolar material in Caenorhabditis are coincident rather than sequential. Accumulation of γ-tubulin at centrioles is mediated by the pericentriolar protein SPD-5.[32] RNAi inactivation of either SPD-5 or γ-tubulin reduced centriole assembly to ~40%-50% of the number expected.[51] Those centrioles that did form failed to elongate to full length and separated prematurely from their parents. Other studies correlated centriole length with the corresponding size of centrosomes; reduced expression of SAS-4, SAS-5 or SAS-6 generated shortened centrioles and reduced the corresponding size of the pericentriolar mass and microtubule aster.[49] These studies suggest that the aster of microtubules formed by centrosomes reflects centriole length and presumably stability. An outstanding question is how recruitment of pericentriolar material (γ-tubulin and SPD-5) is integrated with centriole assembly (ZYG-1, SPD-2, SAS-4, SAS-5, SAS-6) and formation of singlet microtubules.

A widely held perception is that centrioles are effectively immortal; once assembled, centrioles are inherited in successive cell cycles as preformed protein structures. This perception is well founded in that centrioles are known to be highly stable to conditions that disassemble other microtubule arrays.[52,53] However, direct analysis of centriole longevity is limited. Two lines of evidence suggest that centriole structure in Caenorhabditis is dynamic to a limited extent. The first set of studies showed that the level of SAS-4:GFP in preassembled centrioles is reduced by reduced expression of SAS-5 or SAS-6 by RNAi. The simplest explanation of this finding is that SAS-5 and SAS-6 maintain centriole structure and concomitantly, the mass of associated pericentriolar components. The second set of studies used marked mating experiments in which sperm of wild-type males fertilized eggs expressing SAS-6:GFP. The results of this experiment showed that SAS-6 becomes associated with both newly formed centrioles as well as the preassembled parental centrioles. It is not yet known whether SAS-6 is incorporated into the barrel of the preassembled centriole. These studies suggest that centrioles are capable of disassembly, albeit perhaps more so in centrioles with singlet microtubules or lacking δ-,ε-,ζ-and/or ζ-tubulins. In this context it is noteworthy that the absence of basal bodies in sensory cilia in Caenorhabditis[54] could reflect disassembly.

Centriole Inheritance by Mitosis

Centriole Function in Cell Cycle Progression

Centrioles assemble near the proximal end of a preexisting centriole that was built in the previous cell cycle and inherited by mitosis and cell division. Centrioles can form de novo in the absence of a parental centriole, but this pathway is largely restricted to exceptional cells in early development and experimentally applied conditions. The concept of a preformed microtubule template or a centriole-localized genome that directs centriole assembly has been discarded as it lacks experimental support.[55,56] Yet, the nearly universal pattern of assembly suggests that the parental centriole contributes in some way to assembly of the new centriole. One of the first studies to address this question in cultured cells used microsurgical techniques to generate cytoplasts that contained centrioles, but lacked a nucleus and karyoplasts that contained a nucleus, but lacked centrioles (Fig. 5). A central finding was that the karyoplast remained viable, but new centrioles did not form.[57] The underlying reason was not clear until subsequent studies followed cell fragments with continuous microscopic observation.[58,59] These studies produced three important findings. The first finding was that acentriolar karyoplasts first entered and completed mitosis, supporting in vitro studies showing centrosomes are not essential for assembly of bipolar spindles.[60] The second finding was that acentriolar cells successfully completed cytokinesis less often that cells with centrioles, suggesting a role for centrioles in the fidelity of cytokinesis. The third finding was that the acentriolar cells generated by cytokinesis arrested prior to entry into S phase, as if cells cannot exit G1 without a centriole in much the same way that cells cannot exit mitosis without a bipolar spindle.

An outstanding question is whether acentriolar cells arrest in response to the absence of centrioles[61] or whether removal of centrioles elicits a stress driven response that arrests the cell

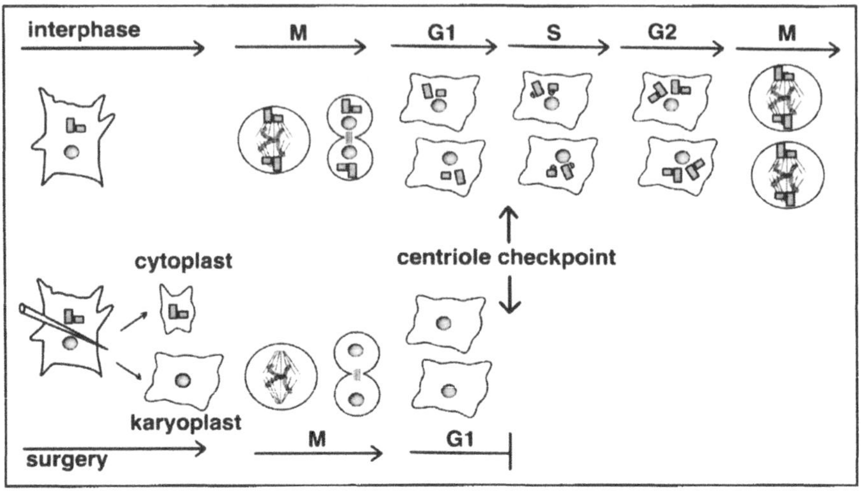

Figure 5. Centriole checkpoint at the G1/S transition. Acentriolar cells were generated by microsurgery with the sharpened tip of glass pipette[58] as shown here or by laser ablation.[59] Karyoplasts complete mitosis and often cytokinesis, arresting before passing the G1/S phase transition. Mitosis in control cells is depicted at the top. For simplicity, asters of microtubules in interphase cells are not shown.

cycle.[62] Although the data is fragmentary and sometimes contradictory,[63] the list of links between the DNA damage response machinery and centrosomes is lengthy,[63,64] including checkpoint kinase-2, p53, BRCA1, BRCA2 and other cell cycle regulators. Among other centrosomal proteins, reduced expression of the pericentriolar components pericentrin and pericentriolar material-1 in vertebrates elicits p53-dependent exit from the cell cycle.[62] These observations suggest that removal of centrioles from cells by microsurgery or laser ablation may have the same effect on the G1/S checkpoint as loss of centrosomal proteins or centrosome organization.

De Novo Assembly of Centrioles

The role of G1/S checkpoint in acentriolar cells was examined in HeLa cells expressing the centriolar protein centrin tagged with GFP.[65] HeLa cells were originally derived from a human adenocarcinoma. HeLa cells, like other transformed cancer cells, show suppressed p53 and Rb function that is needed for surveillance of the G1/S checkpoint. Centrosomes at the poles of metaphase spindles were ablated with a laser and acentriolar cells were generated by cytokinesis. Acentriolar HeLa cells progressed though the cell cycle at the same pace as sibling control cells with centrosomes and, in contrast to normal nontransformed cells, centrioles formed de novo (Fig. 6). One implication of these results is that de novo assembly of centrioles in normal cells is blocked by G1/S checkpoint surveillance. If this view is validated by other tests for the G1/S checkpoint, de novo assembly of centrioles may contribute to centrosome amplification and genome instability that is characteristic of cancer cells.[47,63,64,66]

De novo assembled centrioles in HeLa cells are comparable to centrioles formed while tethered to a parent in two important ways (Fig. 6). First of all, de novo assembled centrioles showed normal structure at the ultrastructural level. This is in contrast to centrioles assembled de novo in cells delayed in S phase, which included a mixture of normal and malformed centrioles.[67,68] Thus, malformed centrioles assembled de novo in S phase arrested cells may, directly or indirectly, reflect delay in S phase or the methods used to elicit S phase delay. Secondly, newly formed centrioles behaved like those in nontransformed cells (Fig. 6). Following de novo

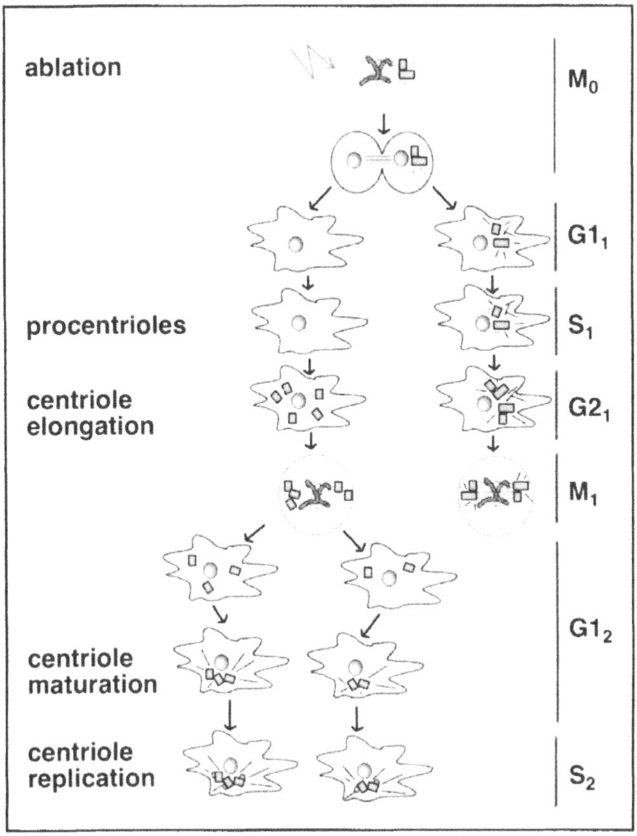

Figure 6. De novo assembly of centrioles in HeLa cells. Following laser ablation of centrioles and completion of mitosis (M_0), centrin-GFP in presumptive procentrioles (small dots) form in the first S-phase (S_1). Pro/centrioles coalesce in $G1_2$ when centrioles acquire a robust pericentriolar mass and an aster of microtubules. Note that immature de novo assembled centrioles are depicted at the poles of a bipolar spindle at M_1 for simplicity, but multipolar spindles were also formed. Because immature centrioles are not associated with microtubules, positioning of immature de novo assembled centrioles at spindle poles in M_1 may reflect a microtubule-independent mechanism. Control cells retaining centrioles undergo normal cell cycle progression, but only G_1-M_1 is depicted here.

assembly, pro/centrioles showed random excursions throughout the cell during the first cell cycle, showing minimal pericentriolar material and virtually no microtubules that could impede movement. Similar excursions of the newly formed centrioles are evident in some cell types where the parental and newly formed centrioles disjoin rather than remaining associated as a centriole pair.[69] In both cases, centrioles coalesced and formed stationary groups that were surrounded by a mass pericentriolar material and a radial array of microtubules.

De novo assembly of centrioles in HeLa cells was delayed in comparison to those assembled in nontransformed cells (Fig. 6). Multiple centrin-GFP aggregates appeared during S phase, but not thereafter, consistent with a process initiated and limited to S-phase as in normal cells. However, centriolar structures were not typically recognized until near the G2/M transition into mitosis rather than in S phase as in nontransformed cells. Newly formed centrioles did not become parents until S_2 after they had acquired a mass of pericentriolar components. Likewise, masses of pericentriolar material appeared in S phase arrested cells before de novo assembly of centrioles. Thus, it is not yet clear whether the accelerated rate of centriole assembly in normal cells reflects factors or structures associated with the preassembled centriole or with the preassembled mass of pericentriolar components.

Proliferation of Acentriolar Cells

Despite the expectation that cell cycle checkpoints will preclude proliferation of cells with defects in centriole assembly, acentriolar exceptions have been described in the literature for some time.[70] The recently reported *DSas-4* mutant in Drosophila is representative of the paradox presented by proliferation of acentriolar cells. The *DSas-4* mutant contains a transposon insertion in a gene encoding a 902 amino acid protein, including a ~ 70 amino acid region with weak homology to SAS-4 in Caenorhabditis and a longer carboxyl terminal ~180 amino acid region with ~45% homology to vertebrate CPAP/CenpJ that is implicated in microcephaly and Tcp10 that is involved in spermatogenesis.[71] Molecular analysis is incomplete and it is not clear whether the insertion mutant is a null or whether DSas-4 is incorporated into centriole barrels. Nonetheless, ultrastructural analysis indicated the absence of identifiable centrioles in somatic cells of these mutants. Spindle assembly was slowed, but mitosis proceeded without significant cell cycle delay and development is remarkably normal, save for aberrant cilia and flagella. Assembly of acentriolar bipolar spindles was ascribed to the default pathway in which microtubules are bundled into bipolar spindles by microtubule dependent motor proteins.

The outstanding question is how acentriolar cells escape cell cycle checkpoints. A plausible explanation for cell proliferation in *DSas-4* Drosophila mutants is that gradual dilution of wild-type protein may circumvent the barriers erected by the abrupt loss of centrosomes by RNAi in nematodes and by microsurgery or laser ablation in cultured cells. DSas-4 deficient mutants die shortly after birth, likely secondary to ciliary defects in sensory neurons. Homozygotes are generated by inbreeding heterozygous mutants. Wild-type protein contributed by heterozygous females to unfertilized eggs sustains embryogenesis, but is gradually lost in sequential mitotic divisions during larval development. Such dilution of DSas-4 may allow mutant cells to gradually adjust to changes in centriole length and/or relocation of pericentriolar material and associated cell cycle regulators. Similar explanations may apply to acentrosomal *centrosomin*[72] and *asterless*[73] mutants in Drosophila, but may not apply to centriole defective mutants in other organisms like the BALD2 and BALD10 deficient Chlamydomonas mutants.[74]

Perspectives

The centrosome is essential for proliferation of virtually all animal cells; in cells born with a centrosome, centriole duplication and separation of centrosomes are prerequisite for spindle bipolarity. This presents a paradox since spindle assembly in acentriolar cell extracts[3] and cell proliferation in mutants lacking centrioles and/or centrosomes[71,74] are compelling arguments that centrosomes are not essential structures.[75] Resolution of this paradox will require answers for several outstanding questions. One outstanding question is how pericentriolar components accumulate at centrioles. Genetic studies suggest centrioles and the pericentriolar matrix have common components and centriole assembly is coincident with recruitment of pericentriolar material, but the molecular mechanisms are not clear. Another unresolved question is how centriole function in organizing the pericentriolar matrix dominates spindle organization. The inability of monopolar spindles to transform into bipolar structures suggests that recruitment of pericentriolar components by centrioles somehow blocks chromatin nucleation of microtubules or that the proteins that nucleate and bundle microtubules are preferentially associated with centrosome organized half-spindles. A third outstanding question is whether the DNA damage machinery is functionally or structurally associated with centrosomes. This is an intriguing possibility given duplication of centrioles and chromosomes in S phase may require common proteins. Fourth, it is not clear whether the essential components of centrioles and centrosomes have been identified or conversely, whether all of the components identified are actually centrosomal proteins. Emerging studies have identified unsuspected players[76] in spindle assembly that will likely impact the role ascribed to centrosomes and centrioles in inheritance. Finally, the potential impact of the experimental methods on centriole inheritance must be considered in evaluating experimental outcomes. While this manuscript was in press, de novo centriole assembly was reported to occur in normal nontransformed cells following laser ablation of centioles,[77] suggesting that cell cycle arrest is some cases reflects laser induced stress. Taken together, the current literature indicates that centriole inheritance reflects self-assembly of centrioles that subsequently dominate microtubule organization and mitosis.

Acknowledgements

I am grateful to the editor and reviewers for critical reading of the manuscript and helpful suggestions. I want to thank Dr. S. Nickerson and the Department of Animal and Dairy Science at the University of Georgia for defraying charges for color figures.

References

1. Moritz KB, Sauer HW. Boveri's contributions to developmental biology—A challenge for today. Int J Dev Biol 1996; 40(1):27-47.
2. Sathananthan AH, Ratnasooriya WD, de Silva A et al. Rediscovering Boveri's centrosome in Ascaris (1888): Its impact on human fertility and development. Reprod Biomed Online 2006; 12(2):254-270.
3. Gadde S, Heald R. Mechanisms and molecules of the mitotic spindle. Curr Biol 2004; 14(18):R797-805.
4. Hannak E, Heald R. Xorbit/CLASP links dynamic microtubules to chromosomes in the Xenopus meiotic spindle. J Cell Biol 2006; 172(1):19-25.
5. Maiato H, Khodjakov A, Rieder CL. Drosophila CLASP is required for the incorporation of microtubule subunits into fluxing kinetochore fibres. Nat Cell Biol 2005; 7(1):42-47.
6. Khodjakov A, Cole RW, Oakley BR et al. Centrosome-independent mitotic spindle formation in vertebrates. Curr Biol 2000; 10(2):59-67.
7. Habedanck R, Stierhof YD, Wilkinson CJ et al. The Polo kinase Plk4 functions in centriole duplication. Nat Cell Biol 2005; 7(11):1140-1146.
8. Bettencourt-Dias M, Rodrigues-Martins A, Carpenter L et al. SAK/PLK4 is required for centriole duplication and flagella development. Curr Biol 2005; 15(24):2199-2207.
9. Verollet C, Colombie N, Daubon T et al. Drosophila melanogaster gamma-TuRC is dispensable for targeting gamma-tubulin to the centrosome and microtubule nucleation. J Cell Biol 2006; 172(4):517-528.
10. Muller H, Fogeron ML, Lehmann V et al. A centrosome-independent role for gamma-TuRC proteins in the spindle assembly checkpoint. Science 2006; 314(5799):654-657.
11. Wilson PG, Fuller MT, Borisy GG. Monastral bipolar spindles: Implications for dynamic centrosome organization. J Cell Sci 1997; 110(Pt 4):451-464.
12. Wilson PG, Simmons R, Saighal S. Novel nuclear defects in KLP61F-deficient mutants in Drosophila are partially suppressed by loss of Ncd function. J Cell Sci 2004; 117(Pt 21):4921-4933.
13. Zariwala MA, Knowles MR, Omran H. Genetic defects in ciliary structure and function. Annu Rev Physiol 2006.
14. Desai A, Mitchison TJ. Microtubule polymerization dynamics. Annu Rev Cell Dev Biol 1997; 13:83-117.
15. Nogales E, Wang HW. Structural mechanisms underlying nucleotide-dependent self-assembly of tubulin and its relatives. Curr Opin Struct Biol 2006; 16(2):221-229.
16. Nogales E, Wang HW. Structural intermediates in microtubule assembly and disassembly: How and why? Curr Opin Cell Biol 2006; 18(2):179-184.
17. Wiese C, Zheng Y. Microtubule nucleation: {Gamma}-tubulin and beyond. J Cell Sci 2006; 119(Pt 20):4143-4153.
18. Moritz M, Agard DA. Gamma-tubulin complexes and microtubule nucleation. Curr Opin Struct Biol 2001; 11(2):174-181.
19. Janson ME, Setty TG, Paoletti A et al. Efficient formation of bipolar microtubule bundles requires microtubule-bound gamma-tubulin complexes. J Cell Biol 2005; 169(2):297-308.
20. Shimamura M, Brown RC, Lemmon BE et al. Gamma-tubulin in basal land plants: Characterization, localization, and implication in the evolution of acentriolar microtubule organizing centers. Plant Cell 2004; 16(1):45-59.

21. Hashimoto T, Kato T. Cortical control of plant microtubules. Curr Opin Plant Biol 2006; 9(1):5-11.
22. Wilson PG, Zheng Y, Oakley CE et al. Differential expression of two gamma-tubulin isoforms during gametogenesis and development in Drosophila. Dev Biol 1997; 184(2):207-221.
23. Martin MA, Osmani SA, Oakley BR. The role of gamma-tubulin in mitotic spindle formation and cell cycle progression in Aspergillus nidulans. J Cell Sci 1997; 110(Pt 5):623-633.
24. Wilson PG, Borisy GG. Maternally expressed gamma Tub37CD in Drosophila is differentially required for female meiosis and embryonic mitosis. Dev Biol 1998; 199(2):273-290.
25. Strome S, Powers J, Dunn M et al. Spindle dynamics and the role of gamma-tubulin in early Caenorhabditis elegans embryos. Mol Biol Cell 2001; 12(6):1751-1764.
26. Hannak E, Oegema K, Kirkham M et al. The kinetically dominant assembly pathway for centrosomal asters in Caenorhabditis elegans is gamma-tubulin dependent. J Cell Biol 2002; 157(4):591-602.
27. Stumpff J, Kellogg DR, Krohne KA et al. Drosophila Weel interacts with members of the gammaTURC and is required for proper mitotic-spindle morphogenesis and positioning. Curr Biol 2005; 15(17):1525-1534.
28. Vogt N, Koch I, Schwarz H et al. The {gamma}TuRC components Grip75 and Grip128 have an essential microtubule-anchoring function in the Drosophila germline. Development 2006.
29. Schnorrer F, Luschnig S, Koch I et al. Gamma-tubulin37C and gamma-tubulin ring complex protein 75 are essential for bicoid RNA localization during drosophila oogenesis. Dev Cell 2002; 3(5):685-696.
30. Khodjakov A, Rieder CL. The sudden recruitment of gamma-tubulin to the centrosome at the onset of mitosis and its dynamic exchange throughout the cell cycle, do not require microtubules. J Cell Biol 1999; 146(3):585-596.
31. Megraw TL, Li K, Kao LR et al. The centrosomin protein is required for centrosome assembly and function during cleavage in Drosophila. Development 1999; 126(13):2829-2839.
32. Hamill DR, Severson AF, Carter JC et al. Centrosome maturation and mitotic spindle assembly in C. elegans require SPD-5, a protein with multiple coiled-coil domains. Dev Cell 2002; 3(5):673-684.
33. Doxsey SJ, Stein P, Evans L et al. Pericentrin, a highly conserved centrosome protein involved in microtubule organization. Cell 1994; 76(4):639-650.
34. Srayko M, Kaya A, Stamford J et al. Identification and characterization of factors required for microtubule growth and nucleation in the early C. elegans embryo. Dev Cell 2005; 9(2):223-236.
35. Popov AV, Severin F, Karsenti E. XMAP215 is required for the microtubule-nucleating activity of centrosomes. Curr Biol 2002; 12(15):1326-1330.
36. Kerssemakers JW, Laura Munteanu E, Laan L et al. Assembly dynamics of microtubules at molecular resolution. Nature 2006.
37. Brittle AL, Ohkura H. Mini spindles, the XMAP215 homologue, suppresses pausing of interphase microtubules in Drosophila. Embo J 2005; 24(7):1387-1396.
38. Al-Bassam J, van Breugel M, Harrison SC et al. Stu2p binds tubulin and undergoes an open-to-closed conformational change. J Cell Biol 2006; 172(7):1009-1022.
39. Calarco-Gillam PD, Siebert MC, Hubble R et al. Centrosome development in early mouse embryos as defined by an autoantibody against pericentriolar material. Cell 1983; 35(3 Pt 2):621-629.
40. Heald R, Tournebize R, Habermann A et al. Spindle assembly in Xenopus egg extracts: Respective roles of centrosomes and microtubule self-organization. J Cell Biol 1997; 138(3):615-628.
41. Stearns T, Evans L, Kirschner M. Gamma-tubulin is a highly conserved component of the centrosome. Cell 1991; 65(5):825-836.
42. Fuller SD, Gowen BE, Reinsch S et al. The core of the mammalian centriole contains gamma-tubulin. Curr Biol 1995; 5(12):1384-1393.
43. Dutcher SK. Long-lost relatives reappear: Identification of new members of the tubulin superfamily. Curr Opin Microbiol 2003; 6(6):634-640.
44. O'Toole ET, Giddings TH, McIntosh JR et al. Three-dimensional organization of basal bodies from wild-type and delta-tubulin deletion strains of Chlamydomonas reinhardtii. Mol Biol Cell 2003; 14(7):2999-3012.
45. Chang P, Giddings Jr TH, Winey M et al. Epsilon-tubulin is required for centriole duplication and microtubule organization. Nat Cell Biol 2003; 5(1):71-76.
46. Quarmby LM, Mahjoub MR. Caught Nek-ing: Cilia and centrioles. J Cell Sci 2005; 118(Pt 22):5161-5169.
47. Swallow CJ, Ko MA, Siddiqui NU et al. Sak/Plk4 and mitotic fidelity. Oncogene 2005; 24(2):306-312.
48. Ishikawa H, Kubo A, Tsukita S et al. Odf2-deficient mother centrioles lack distal/subdistal appendages and the ability to generate primary cilia. Nat Cell Biol 2005; 7(5):517-524.
49. Leidel S, Gonczy P. Centrosome duplication and nematodes: Recent insights from an old relationship. Dev Cell 2005; 9(3):317-325.

50. Delattre M, Canard C, Gonczy P. Sequential protein recruitment in C. elegans centriole formation. Curr Biol 2006; 16(18):1844-1849.
51. Dammermann A, Muller-Reichert T, Pelletier L et al. Centriole assembly requires both centriolar and pericentriolar material proteins. Dev Cell 2004; 7(6):815-829.
52. Bobinnec Y, Khodjakov A, Mir LM et al. Centriole disassembly in vivo and its effect on centrosome structure and function in vertebrate cells. J Cell Biol 1998; 143(6):1575-1589.
53. Rosenbaum J. Cytoskeleton: Functions for tubulin modifications at last. Curr Biol 2000; 10(21):R801-803.
54. Wright KA. Peripheral sensilla of some lower invertebrates: The platyhelminthes and nematoda. Microsc Res Tech 1992; 22(3):285-297.
55. Fulton C. Origin and continuity of cell organelles. New York: Springer-Verlag, New York Inc., 1971.
56. Marshall WF, Rosenbaum JL. Are there nucleic acids in the centrosome? Curr Top Dev Biol 2000; 49:187-205.
57. Maniotis A, Schliwa M. Microsurgical removal of centrosomes blocks cell reproduction and centriole generation in BSC-1 cells. Cell 1991; 67(3):495-504.
58. Hinchcliffe EH, Miller FJ, Cham M et al. Requirement of a centrosomal activity for cell cycle progression through G1 into S phase. Science 2001; 291(5508):1547-1550.
59. Khodjakov A, Rieder CL. Centrosomes enhance the fidelity of cytokinesis in vertebrates and are required for cell cycle progression. J Cell Biol 2001; 153(1):237-242.
60. Wadsworth P, Khodjakov A. E pluribus unum: Towards a universal mechanism for spindle assembly. Trends Cell Biol 2004; 14(8):413-419.
61. Murray AW. Cell cycle: Centrioles at the checkpoint. Science 2001; 291(5508):1499-1502.
62. Srsen V, Gnadt N, Dammermann A et al. Inhibition of centrosome protein assembly leads to p53-dependent exit from the cell cycle. J Cell Biol 2006; 174(5):625-630.
63. Loffler H, Lukas J, Bartek J et al. Structure meets function—Centrosomes, genome maintenance and the DNA damage response. Exp Cell Res 2006; 312(14):2633-2640.
64. Srsen V, Merdes A. The centrosome and cell proliferation. Cell Div 2006; 1:26.
65. La Terra S, English CN, Hergert P et al. The de novo centriole assembly pathway in HeLa cells: Cell cycle progression and centriole assembly/maturation. J Cell Biol 2005; 168(5):713-722.
66. Fukasawa K. Centrosome amplification, chromosome instability and cancer development. Cancer Lett 2005; 230(1):6-19.
67. Marshall WF, Vucica Y, Rosenbaum JL. Kinetics and regulation of de novo centriole assembly: Implications for the mechanism of centriole duplication. Curr Biol 2001; 11(5):308-317.
68. Khodjakov A, Rieder CL, Sluder G et al. De novo formation of centrosomes in vertebrate cells arrested during S phase. J Cell Biol 2002; 158:1171-1181.
69. Piehl M, Tulu US, Wadsworth P et al. Centrosome maturation: Measurement of microtubule nucleation throughout the cell cycle by using GFP-tagged EB1. Proc Natl Acad Sci USA 2004; 101(6):1584-1588.
70. Feldman JL, Marshall WF. Centrioles: Bad to be bald? Curr Biol 2004; 14(16):R659-660.
71. Basto R, Lau J, Vinogradova T et al. Flies without centrioles. Cell 2006; 125(7):1375-1386.
72. Megraw TL, Kao LR, Kaufman TC. Zygotic development without functional mitotic centrosomes. Curr Biol 2001; 11(2):116-120.
73. Giansanti MG, Gatti M, Bonaccorsi S. The role of centrosomes and astral microtubules during asymmetric division of Drosophila neuroblasts. Development 2001; 128(7):1137-1145.
74. Dutcher SK. Elucidation of basal body and centriole functions in Chlamydomonas reinhardtii. Traffic 2003; 4(7):443-451.
75. Marshall WF. What is the function of centrioles? J Cell Biochem 2006.
76. Schreiber V, Dantzer F, Ame JC et al. Poly(ADP-ribose): Novel functions for an old molecule. Nat Rev Mol Cell Biol 2006; 7(7):517-528.
77. Uetake Y, Loncarek J, Nordberg JJ et al. Cell cycle progression and de novo centriole assembly after centrosomal removal in untransformed human cells. J Cell Biol 2007; 176(2):173-182.

Index

I

J

K

M

N

P

Q

R

S